Reality. Not Metaphysics

Robert Innes

I am now convinced theoretical physics is actually philosophy.
Max Born

Consciousness cannot be accounted for in physical terms.
For consciousness is absolutely fundamental.
It cannot be accounted for in terms of anything else.
Erwin Schrödinger

. . .the world of our sense experiences is comprehensible. The fact that it is comprehensible is a miracle.
Albert Einstein

Looking at candy can be sweeter than chewing it
Lulu Harris

Dedicated to the Memory of
Arthur F. Setteducati

Copyright © 2016 Robert Innes

All rights reserved

First Edition - September 2016

Printed by CreateSpace, an Amazon.com Company

ISBN13:9781541156920

ISBN10:1541156927

Preface

The purpose of this book is to show that bewildering paradoxes of the quantum mechanics and relativity that burden our understanding of reality, arise from metaphysical appendages to physics that subconsciously protect our preconception of what is real. This metaphysical burden falls away when we recognize reality as consciousness: Because we all experience consciousness, it is not difficult to know it is real. This reality does not come from belief, but transcends anything we may believe. What we can believe we will define as awareness and show that it is very different from consciousness. Both are commonly confused because they are always concomitant. This confusion blinds one to understanding the role of consciousness in the quantum mechanics: We are not exhuming debunked interpretations of the quantum mechanics that ascribe wave-function collapse to observation entering "living consciousness" or "mind".

What we have to give up is difficult: Believing in the ultimate reality of the physical world: It is hardwired by evolution into our neural anatomy. Once this detachment is made, the conception of reality becomes very different and almost obvious. The difficulty of doing this is because we create reality in the image of our theories about it. Almost a century of the history of the quantum mechanics is one of the evolving entrapment of its principle architects by unwary models of reality, while disavowing intellectually that it is the task of science to talk about reality at all. Einstein did talk about reality, but made the mistake of placing his faith in physical reality. A parallel we have gone through in the past is giving up believing that the world is flat. It was very disturbing to those who had to do it, but the pain went away quickly once the truth dawned on them.

It is not meaningful to discuss metaphysics outside the context of physics.

One can understand the gist of the Special Theory of Relativity without going into mathematical details, provided the reader is willing to accept certain scientifically accepted conclusions without proof. Pointer readings to scales and clocks in one system are related to similar readings in another system where the two systems are in relative motion. The relationship between pointer readings had always been the basis of mathematical theory in exact sciences. Relativity made some changes in the equations dating from antiquity, which related these two sets of readings. Even though the math is basic high school algebra, it can be omitted or relegated to an appendix for interested readers who otherwise do not know it.

However, the quantum mechanics is not as easy. My purpose is to clarify why widely accepted beliefs concerning the quantum mechanics do not follow from the mathematics: They are metaphysical constructs protecting intuitive prejudice which not only fails to reveal truth but instead obscures it with pseudo-truth. One needs a sufficient understanding of the mathematics to see why, as Hugh Everett showed, it supplies its own interpretation. The quantum mechanics is not a relationship between pointer readings: Observables in the quantum mechanics are the eigenstates of the observing operator, or more exactly, their eigenvalues: At first, this mathematics was over the heads of even the founding fathers of the quantum mechanics. If the reader has no knowledge of mathematics beyond what they remember after forgetting high school algebra, I believe that quantum math can be understood well enough to see how the metaphysical purge is possible. Readers with no knowledge of algebra can skip the math, but will have to accept on faith that quantum math supplies its own interpretation. So as not to drag readers through math and physics with which they are familiar, I have put details in appendices. The appendixes are not intended to teach the reader math and physics, but to demystify them to the extent that a layman can see the role they play in theory. If one reads this book more than once, the appendices, if needed at all, can be omitted on the first pass.

A famous author, I forget who, apologized for the length of his presentation saying he could have made it shorter it if he had had more time. Without trying to be funny, I make the same excuse.

Robert Innes September 2016
Sebastopol, California
rinnes@mcn.org

Table of Contents

Introduction..8
Semantics..16
Paradox..18
History...19
   Pre-quantum history..20
   A brief history of the quantum mechanics......................23
   The funeral of Newtonian Mechanics.............................25
   The Quantum Mechanics departs from earlier science.....28
   Brief history of the theories of relativity............................28
   What Einstein did..29
   Lorenz Transformations...33
   Measurement of space and time.....................................34
   Coriolis force...37
   The place of classical physics...37
Quantum Mechanics..38
   Wave functions..38
   What Heisenberg did..39
   What Schrödinger did...42
   The Amplitudes of Heisenberg and Schrödinger.............44
   The quantum mechanics makes her debut....................45
   What Born did..45
   Wave interference...46
   Particles interfere with themselves. Paradox?.................48
   Quantum amplitudes..48
   Equivalence of Schrödinger and Heisenberg..................50
   Hilbert Space..54
   What Dirac did...54
   The rules of the quantum amplitude game.....................56
   Eigenfunctions of an operator...57
   Observations in the quantum mechanics........................60
   Feynman on amplitudes, circa 1960...............................67

Metaphysics: Interpretations of the Quantum Mechanics....70
    Copenhagen Interpretation, 1927..70
    "EPR" Interpretation, 1935...71
    John Bell throws a wrench into the EPR works..................73
    What John Bell did..74
    EPR proven wrong..75
    Bohm - de Broglie Interpretation:..76
    The Relative States Interpretation, 1957...............................76
    The Basis of Everett's formulation..78
    The Many Minds Interpretation..80
    Feynman and metaphysics..80
    Remarks on Feynman's view..81
    Leonard Mandel, 1992..82
    The Fall of Classical Physics...83
    Intuition and common sense...85
Consciousness..86
    The Doctrine of Psycho-Physical Parallelism......................86
    The reality of consciousness...88
    The non-existence of consciousness......................................88
    George Berkeley..89
    Schrödinger..91
    The paradox of the Samkhya philosophers.........................91
    Principle of Vedanta...93
    Michael Gazzaniga...93
    Gazzaniga and the Principle of Vedanta..............................94
    Awareness is the binding of consciousness........................96
    Schrödinger and Gazzaniga after brain surgery.................96
    You in consciousness and you in awareness......................98
    The Samkhya paradox and you...99
    The binding of consciousness thought "experiment".....100
    Consciousness binding experiments you do not need....101
    Consciousness in antiquity...103
    Descartes..103
    Consciousness in the modern world...................................104
    Phantom limbs..106

- The Honey bee..107
- Do computers experience consciousness?........................107
- Hugh and John on psycho-physical parallelism..............112
- Gazzaniga returns..113
- Consciousness - Awareness; Reality - Existence..................115
  - Consciousness..118
  - Awareness..118
  - Consciousness vs Awareness...120
  - Awareness of Consciousness...120
  - Roger Penrose..121
  - Reality and Existence...122
  - Creating reality in the image of theory............................124
  - Elementary particles..124
  - Reality and Science..127
  - Three true stories...128
  - The axioms of consciousness..130
  - The properties of awareness...132
  - The Brain...134
  - Observation: Classical vs Quantum Mechanics...............136
- The Uncertainty Principle..144
  - The "Uncertainty Principle" and light.............................151
  - How God created elementary particles............................155
- The Foundation of Mathematics..156
  - Summary of the history of mathematical logic................156
  - Gödel's Proof...157
  - Meta-mathematics..158
  - Russell, Gödel, Grade school...160
  - Elementary particles are mathematically exact................161
- Relativity...163
- Quantum Mathematics Interprets Itself...............................164
  - How can we escape these metaphysical contortions?......168
  - Understanding quantum cause and effect........................169
  - Coherence time..176
  - Astronomical telescope: An anti-interferometer.............177
  - What Hugh Everett gave us:..180

- The Einstein, Podolsky, Rosen Experiment..........184
  - Quantum entanglement..........184
- Leonard Mandel..........187
  - Mandel's interferometer..........188
  - Mandel Found:..........192
  - Look Ma, no wires, no cables, no pull strings!..........193
  - Slogging out of the metaphysical quagmire..........196
  - The magic of quantum non-locality..........198
  - Causal inconsistency..........198
  - Knowing "which path" destroys interference..........199
  - Seeing interference in Mandel's interferometer..........199
  - Understanding what Alice and Bob see..........201
  - Banning..........203
- Electron Spin..........205
  - Stern-Gerlach Experiment..........205
  - Pauli spin matrices..........208
- Arrow of Time..........217
  - Lost in an infinite realm?..........221
  - Free will..........221
- Summary..........226
  - Physical Objects..........226
  - The Physical World and the quantum Mechanics..........229
  - Feynman diagrams..........232
  - EPR and Everett..........233
  - The frying-pan..........234
  - Niels Bohr..........236
  - It and you..........236
  - Philosophy of life..........238
  - What happened to Schrödinger's cat..........240
  - Evolution and the Uncertainty Principle..........242
- Conclusion..........246
  - The Physical World..........246
  - How did God create the Physical Universe?..........246
  - Evolution and Creationism..........250
  - Why consciousness now?..........251

The Continuum..253
In the beginning..254
Belief and Truth...255
Parable of Physical Reality..257
The dreams of reality and the reality of dreams.............257
God...257
Appendix A: Technical terms...259
Appendix B: Lorentz Transformations..................................263
Appendix C: Infinitesimal Calculus..264
   Functions..264
   Rates of change of functions..264
      More than one independent variable...........................267
   The exponential function..267
Appendix D: $\sqrt{-1}$ makes waves..269
   Functions of a complex variable.......................................270
Appendix E: Linear Algebra..272
   Linear Algebra..272
   Vectors..272
   Scalar products...273
   Matrices..275
Appendix F: Differential Equations...277
   Linear operators...277
   Demystifying differential equations................................278
   Eigenfunctions..284
   Eigenfunctions are orthonormal.......................................284
   Linear superposition of eigenfunctions..........................285
   Hilbert Space and the quantum mechanics....................286
Appendix G: The Twin Paradox of the Special Theory.......287
Appendix H: The Cast..292

Introduction

From the time of Galileo and Newton to the end of the 19th century was the greatest era of advancement in all prior history of the understanding of experience. It made inroads into such diverse fields of endeavor as the chemistry of life; engineering; astronomy. It appeared that all the fundamental theories of science were known. Much remained to be explored, but there was no reason to belief that new discovery would not be supported by then known fundamental theories of geometry, dynamics, and the theories of fundamental forces of gravity, electricity and magnetism. This is now called the classical era of science.

No one would have guessed that the greatest era of science was about to begin. When it began, it was not without pain. It was heralded by two paradoxes:

- A precise and simple experiment by Prof. Albert Michelson at what is now the Case Western Reserve University in Ohio showed that the earth, always, was motionless in space. Even if it were momentarily motionless, its orbital motion around the sun combined with the sun's motion would generally be many tens of kilometers per second or much more. This paradox was resolved by Albert Einstein in 1905 when he showed that mass, length and time, the holy trinity of scientific and civil metrology, were not absolute attributes of physical objects. Ten years later he showed the the hallowed geometry of Euclid, upheld by Immanuel Kant as the correspondence between the mind of man and the mind of God, was wrong. These were Einstein's Special and General theories of Relativity.

- An even more shocking paradox arose out of the theory of radiation from incandescent objects. Two theories were combined: 1. Thermodynamics, highly successful in engineering, based on Newtonian dynamics. 2. A new axiomatic theory, uniting all earlier electromagnetic theory which explained light as electromagnetic waves. 1 and 2 combined not only made predictions that did not match experimental data, but made a prediction so absurd it was described as catastrophe: A red hot poker would emit infinite radiation. Max Planck resolved this paradox in 1900 by recognizing that electromagnetic waves of frequency f, conveyed energy E in packets or quanta, now called photons, each of energy proportional to f thus: $E = h.f$. This was the birth of quantum mechanics.

Planck's theory not only removed the catastrophe, but if the constant h had a particular but very small value, the theory matched experimental data exactly. Now called Planck's constant, h has taken its place beside the velocity of light in the hall of fame of physical constants. But Planck was not a happy camper. He could not bring himself to believe that waves delivered energy in lumps.

Interference of waves seen in water waves and sound waves was also seen in light in 1800. If a wave is split into two beams which later recombine at a point, if the difference the two beams have traveled to get there is a multiple of the wave length, the crests will combine with crests and the troughs with troughs to make a large wave motion. At a different point if a crest of one wave combined with a trough of another, the waves can cancel completely. This effect is called interference. Classical physics presented no problem explaining interference in water, sound or light. In light it was seen as dark and bright fringes. A device showing this effect is called an *interferometer*.

Suppose a beam of light is split along two paths with a half silvered mirror which reflects half of the light and transmits the other half. What does the mirror do to one photon? A reasonable theory would be that the photon is splits into two half photons each with energy of $1/2\,E$. Later these recombine to make a whole photon of energy E in a bright fringe, or cancel to zero in a dark fringe.

This is not what happens.

If detectors are put in both paths, the whole photon is seen in one path or the other but never both. The choice of which path seems to be completely random. Yet when the detectors are removed, interference is seen even if the photons arrive individually.

An even more perplexing paradox arises if two photons are generated by a common event. If one goes through vertically polarized sunglasses the other will be stopped by vertically polarized sunglasses, and conversely. They will make one choice or the other, seemingly at random. Since this will be true even if the sunglasses are light years apart, how does one photon know what the other did? This paradox caused Einstein and his colleagues to disavow the quantum mechanics as a complete theory. Einsteins alternative theory that the photon pair was born with a hidden disposition to appear as they were seen was proven wrong experimentally in the 1980s.

These paradoxes caused the founding fathers of the quantum mechanics to fragment into dissenting factions.

More baffling yet were experiments done by Prof. Leonard Mandel at the University of Rochester circa 1990. He combined both of the above by putting special crystals in each interferometer path. These crystals turned an incoming photon into two outgoing photons. A photon from each crystal was sent to an interference screen which produced interference. Call this part of the apparatus the "signal" interferometer. The other two photons from each crystal went to a second interferometer called the "idler" which was not connected to the "signal" device except both got their photon from the same two crystals. The "idler" interferometer could be in a separate sealed room separated from "signal" by concrete wall, except for two small holes for the "idler" photons to get through. If you block one path of the idler interferometer it stops interference at the idler screen because there is no longer reinforcement and cancellation of the photon waves: Only half the light is reaching the "idler" screen. All the light from both paths is still reaching the "signal' screen in the other room. Humdinger: The interference fringes disappear there too. How can your just knowing what is happening somewhere change what is happening somewhere else when there is no physical connection between what you are doing and what is happening there?

Thirty years before the above experiment, Prof. Richard Feynman of Cal Tech effectively told his undergraduate students that if you can know which path photons are taking in an interferometer, there will be no interference. Prof. Mandel showed that you can without any physical connection to the "signal" interferometer find out which path its photon took by looking in the "idler" path because they both got their photon from the same crystal. The implication of this was known in the earliest days of the quantum mechanics. Einstein refused to accept it to his dying day, calling it "spooky". Had he lived to see Mandel's experiment, he would have been forced to accept it. The history of the quantum philosophy then may have been very different.

The purpose of this epistle is to show that these paradoxes arise from belief in the ultimate reality of the physical world. We will show that they go away when ultimate reality is seen as consciousness.

Experience never lies to us. Our beliefs about it do. We have no contact with reality except consciousness. Without it, whatever we may believe exists would not be manifest. Not thinking about consciousness, we look through it at what we see as reality: the physical world beyond. If our attention is drawn back to consciousness, we brush it off as some spin-off of brain function: the place of music, dreams, thoughts. Until the dawn of the 20th century, every advance in science lent support to the view that the physical world was ultimate reality. Obviously, it was not a dream or a passing thought.

Early in the 20th century, the founding fathers of the quantum mechanics and the theories of relativity were in contentious disagreement about what the word *reality* meant. They did not dispute the representation of observed data by the mathematics of the new theories of physics, but made ingenious constructs to protect their metaphysical views. These were not drawn from some creed of The Order of Metaphysical Fellows, but gut-level intuition by everyone who saw the results of quantum mechanical experiments. One grabs at any straw that allows one to hang on to preconception of what one thinks is real. These reactions derive from central nervous system programs which always have been consistent with daily experience: they are the creators of the "physical word". For the first time, experience of the "physical world" confounded us with paradox. Dozens of brilliant people, working over a quarter century to give us the greatest theory in the history of science, did not celebrate the interpretation of what they had accomplished. Instead, dissenting, they diverged along different metaphysical paths.

A quarter century after the quantum mechanics matured, it was shown by Hugh Everett that the mathematics of the quantum mechanics supplies its own interpretation. The prevailing metaphysical view called the *Copenhagen Interpretation* required two constructs: *The principle of uncertainty; The collapse of state vectors.* Everett showed that the Copenhagen interpretations of random determinism and the need to collapse states were not physics, but metaphysics supporting the belief that a record of a chain of experienced quantum events describes the only reality. This was true of classical physics: any other sequence would violate physical law. In the quantum mechanics, the conservation laws of physics and their associated symmetries remain valid when what one observes is caused by each of a number of eigenstates of the observation: No matter what chain of consequent quantum events happens, each element of the chain will always seem to happen in a "physical world" of classical physics. Everett showed that quantum mechanical mathematics gave no reason to assign a greater reality to the experience caused by one eigenstate of the observation than to any other. This raised a serious objection to Everett's *"Relative State" Formulation* as he called it, that multiple, possibly infinite numbers of physical universes may be incurred by the observation of a single quantum event.

The words introduced below in *bold italics* below will be exactly defined in later text and must be distinguished.

It was recognized by John von Neumann, the mathematical mentor of the quantum mechanics, and later by Everett, that physical theory describes either the physical world, or *consciousness*, in consequence of the principle of psycho-physical parallelism. We show that this metaphysics is removed from the quantum mechanics by recognizing that *reality* is not an imagined "physical world" but *consciousness*. Doing so does not introduce new metaphysics because *consciousness* is not a belief; its *reality* is incontrovertible; it does not require the creation of new energy for each experienced world because consciousness, although *real*, does not *exist* in a physical sense. The proof of *reality* of *consciousness* is its experience: otherwise it <u>is not scientifically demonstrable.</u>

I will use the word *awareness* to describe the record of a chain of observed events caused by the orthogonal eigenstates of the observation. So defined, *awareness* <u>is scientifically demonstrable.</u> Notwithstanding stark distinction between *consciousness* and *awareness* in this context, these words are often used interchangeably: After entering our minds, the experience of awareness is always manifest in consciousness. However one is seldom aware of this duality, blurring the distinction between them. Failing to make this distinction makes understanding of quantum mechanics in the light of consciousness impossible. Making the distinction requires the resolution of a paradox concerning *consciousness*. This paradox, originating in antiquity, was discussed by Erwin Schrödinger in the context of science.

There is no scientific bases for believing that *consciousness* uniquely relates to brain function alone. In Schrödinger's words: *Consciousness cannot be accounted for in physical terms. For consciousness is absolutely fundamental. It cannot be accounted for in terms of anything else.* However, *awareness* of *consciousness* may not be possible without higher brain function.

It is impossible to investigate *consciousness* as a phenomenon like electricity or magnetism were after Newton's time because, unlike these phenomena, *consciousness* does not *exist*. However, it is scientifically possible to relate objective descriptions of a subject's neural fining patterns to the subject's report of subjective experience. This can be extended to objective understanding of electronic device cybernetics to subjective experience if and when such machinery is used to replace defective neurology. These investigations are the basis for future research of *consciousness*. They would answer questions about the cybernetic distinction between *conscious* experience of color vision, of sound, of sensation and so on. They could be done by blind and deaf scientists who can know they succeeded when <u>you</u> tell them that <u>you</u>

consciously can see normally again, without on their part ever knowing what it felt like to see or hear anything. What they will reveal cannot be found by philosophical contemplation, however valuable such contemplation may be in organizing understanding. First, it is necessary to lay the scientific foundation upon which this science will stand. The purpose of this missive is to proclaim that this foundation is a fait accompli: It is the quantum mechanics. The proof of the reality of *consciousness* is you. One cannot prove to you that your *conscious* experience rests on a quantum mechanical foundation, but one can show you the mess you will get into, which is the mess science has gotten into, by believing otherwise.

The elementary particles demonstrably *exist* in some sense: They have exact mathematical properties. Do they stand on the foundation of the world of classical physics, or do they grounded in mathematical truth? The quantum mechanics is the formalism supporting the understanding of everything from atomic spectra to transistors to DNA molecules to elementary particle events produced by particle collisions in the Large Hadron Collider at CERN near Geneva. This formalism is not consistent with views of the classical realm of physical "reality".

We explain that the processes von Neumann described as:

I A discontinuous change brought about by observation of a state that changes it to one of its eigenstates

II A continuous deterministic change of an isolated system

which, seen as events in the physical world, incur paradox from which escape is sought by incurring the breakdown of causality and positing a doctrine of uncertainty of effect. (***Mathematical Foundations of Quantum Mechanics***, John von Neumann. See p351 of the English translation of Beyer)

We associate the word *awareness* with von Neumann's Process I, and *consciousness* with Process II. There is no need for the hypotheses of the physical world. We will show that *awareness* is the binding of *consciousness*. Everett showed that von Neumann's claim that Process I is essential not to be correct. Only Process II is essential and within it nothing is random. He showed that everything needed to explain *awareness*, and he used that word, follows from the mathematics of Process II. In his thesis he does not discuss *reality*, but in his earlier paper alludes to *consciousness* by describing his formulation of the quantum mechanics as not doing violence to the doctrine of psycho-physical parallelism.

Interpreted in this way, the structure of *consciousness* does not have its origin in "the physical world" as von Neumann saw it but, presently as best can be seen, is understandable by the quantum mechanical world of elementary particles which in turn has a foundation in mathematics. The foundation of mathematics was elucidated by Gödel, who showed that it cannot stand on a demonstrably consistent finite foundation of logical axioms representable as physical objects.

In a public lecture in 1964, available in online video, Richard Feynman said "Nobody understands the quantum mechanics". With the benefit of half a century of hindsight, his statement can be strengthened and clarified: "It is impossible to understand the quantum mechanics, if one believes in the ultimate reality of the physical world". The Hilbert Space math of the quantum mechanics represents experience. Provided one can rise above naive beliefs about consciousness and see the distinction between consciousness and awareness, it is not at all difficult to understand how quantum math represents conscious experience. Doing so does not require proficiency in computing quantum amplitudes. Understanding Newton's simple second law differential equation does not require proficiency in solving it. This metaphysical purge of quantum mechanics removes its mind boggling paradoxes. It leaves a simple, almost obvious interpretation: one that is very different from that based on the physical world model. Making this change is not made difficult by accepting consciousness as *reality*, but relinquishing "the physical world" world as *reality*. It not only gives us a very different view of external reality, but us a very different view of ourselves as beings of mortal awareness and immortal consciousness. This transition was not brought about because the quantum mechanics was more accurate than classical mechanics. Commonly used quantum math is a known approximation. It was brought about because the quantum mechanics brought a fundamental change in the structure of scientific theory. This transition was not even true of the theories of relativity, despite the Special Theory being the womb of quantum mechanics of fermions in showing us their wave nature.

A model of physical *reality*, largely consistent with classical physics, is programmed into central nervous systems of sentient beings by effectively making the approximations that the velocity of light is infinite and Planck's constant zero. For survival needs, these approximations are very accurate. This model is an illusion in awareness from which we are not delivered by intellectual knowledge of the correct values of these physical constants. The corresponding states of *consciousness* make us feel intuitively that we are in a "physical world" consistent with the laws of classical physics.

In the same historical period of the emergence of the quantum mechanics, the Special and General theories of relativity emerged quite independently. However, the converse was not true: The Special theory played an important role in the foundation of the quantum mechanics. Both of the theories of relativity, particularly the Special, do considerable violence, manifest as paradox, to the concept of physical reality.

For ninety years, students of physics have been taught in the major universities that the quantum mechanics predicts, for a given experiment, that one of a set of possible events will happen. If the set has only member, that event is certain. If the set has more than one member, which can be infinitely many, which event experienced is unpredictable, not because of lack of sufficient information, but even by God. Superluminal causality "collapses" the other members, which could happen in remote places. This teaching is an artifact of the belief that *reality* is the physical world: It is not physics, but metaphysics. These bizarre absurdities, called the Copenhagen Interpretation, are not needed at all to use the quantum mechanics. Out of one side of their mouths physicists disavow the need to talk about reality at all. Out of the other side, because they believe there is some reality physics must be talking about, they proclaim this metaphysics as fact.

The words *exist* and *existence* have several meanings, at least in English in which I write. We think of a rock as *existing*. Solutions to mathematical equations may be said to *exist*. Numbers may be said to *exist* mathematically and numerical properties of a number may be demonstrated by playing with some number of *existential* beads. However, numbers do not derive their *existential* properties from beads which we think of as *existing* physically. We cannot prove $2^{57,885,161} - 1$ is a prime number by playing with beads, but we can comprehend that it *exists* as a prime. The purpose of this missive is to emphasize that the homeland of beads and rocks and classical mechanics: the physical world, is <u>not</u> reality. Pulling away the veil of physical world reveals Absolute Nothingness: the Void. It is the substance out of which consciousness is made. But the world of our sense experiences is not void. How can physics account for it? It can as a theory, not about physical reality, but the relationship of awareness to consciousness:

Consciousness *cannot be accounted for in physical terms. It cannot be accounted for in terms of anything else.* I didn't say that, Schrödinger did. His words can only describe a miracle.

Awareness, *the world of our sense experiences is comprehensible. The fact that it is comprehensible is a miracle.* I didn't say that. Einstein did.

Semantics

For the purpose of this paper, the four words *consciousness, awareness, reality* and *existence* are to be understood <u>only</u> in the sense in which they are defined here. To the extent that the definitions to be given here are clear, any disagreement with them is therefore semantic. I could have used nonsense codes for these four words and let the semanticists replace them later with whatever they thought appropriate, but I chose not to be so stilted. I wrote in English which is poorly equipped to make distinctions between these words. When we use the word *conscious* used in statement *the patient was unconscious,* we are describing something observable. Feeling dizzy is not observable unless you experience it, but you would be demonstrably *unaware* of anything after blacking out. The word could mean interrupted sensation or it could mean ignorance: *I was aware they are married but unaware they were cousins.* Reality is commonly used to describe the correctness of theory or idea: *His ideas are out of touch with reality* instead of saying in a more clumsy way that they are not demonstrable or do not represent experience. I read of *western consciousness* which is not at all the sense in which the word *consciousness* is used in this book. *Western perception* might be a better choice in this book if I had to refer to it.

Notwithstanding this confusion, please put up with my English choices while reading this book. If you are reading a translation of it into some other language and you find the semantics inappropriate, please put up with the translator's choices and argue about semantics later. It is not my purpose to make semantic choices, but I need four words: Any choice I make can be questioned for good semantic reason..

This paper does not purport to explain consciousness in terms psychological, physical, chemical, anatomical, cybernetic, or any derivative science. It is not seen as "phenomenon" or "mind". The role of consciousness in quantum mechanics has been considered by many from Eugene Wigner and Wolfgang Pauli, pioneers of quantum mechanics, to Roger Penrose in recent times. Michael B. Mensky, Evgeny Ivanov and others discuss consciousness as it relates to Everett's thesis. The premise of consciousness, given by Schrödinger on the dedication page of this book, is discussed by Peter Russell and others. The purpose of this paper is not to build on such investigations, however valid they may be, but to show that the paradoxes of quantum mechanics disappear when the presumed reality of the "physical world" is relinquished. In the light of the principle of psycho-physical parallelism the only reality that remains is consciousness. Reality is not bestowed on consciousness by metaphysical belief. Consciousness is real no matter what we believe. However, in order to understand the quantum mechanics in this light, it is necessary <u>not</u> to confuse the words *consciousness* and *awareness.* It is necessary to resolve paradox that easily can snare anyone thinking about consciousness.

Paradox

Paradox beholders never find:
Paradox beheld is in the mind

Throughout history people have been confounded by paradox: When experience defies belief. It has been a recurring theme in mathematics and physics and in attempts to understand consciousness. The following statement is take here as axiomatic: **Paradox <u>always</u> is caused by flaws in our thinking.** It is not inherent in the origin of experience. There is a price to be paid if one is willing to live with paradox: Doing so accepts a state of illusion, worthwhile only if we paid a stage magician to put us in that state. Knowing the magician's trick spoils the fun.

Since antiquity, the Physical World has been the unshakable rock of ages: The Physical World is consistent: Never before the dawn of the 20th century was physical reality in two contradictory states at the same time. By 1930, the unshakable rock was in shards. Paradox: An indivisible particle could be in two different places at the same time, and they can be a million light years apart. Even more shocking: It can be is superposition of states of existing and not existing at the same time. Before the quantum mechanics, such proclamation would have been considered the babble of lunacy. Ninety years since the quantum mechanics was born, a shard of the dead god, shrouded in metaphysical armor of random and superluminal causality, remains on the altar of worship by most physicists, not because they are heathen, but because they have nothing else to believe in. My purpose is to give reason to believe that there is no physical world nor ever was. I will fail if I cannot give reason to believe that there is a better place to ground ones understanding of the consistency of experience.

History

The classical era of science

Standing on the foundation of the ancient theory of geometry of Euclid, science began an explosive three-century period of development from the time of Galileo and Newton to the end of the 19th century. Grounded in accurate theories of physics, understanding spread across chemistry, astronomy, engineering, geology and other earth sciences, botany, zoology, anatomy, physiology, paleontology and micro-biology, which spurred great advances in mathematics and related sciences like statistics and logic. At the end of that period it appeared that all of the basic theories of reality were understood; the future of physics lay only in determining physical constants to greater accuracy. The machinery of physical law, mathematics, was in the process of being grounded in a demonstrably self-consistent logic, freeing it from proof floundering in ineffable reasoning.

However, late in the 19th century, two dark clouds were seen on the horizon of physics. One was empirical, the other was theoretical:

- An exact experiment to measure the velocity of the earth through space failed, yielding a null result.
- The union of Newtonian physics with a new axiomatic theory that unified all prior electromagnetic theory, made predictions so absurd they were described by physicists as catastrophe.

These dark clouds brought storms that left the classical conception of reality in ruin.

The greatest era of science was yet to begin. By the end of the first third of the 20th century new foundations for scientific theory were laid: The quantum mechanics and the theories of relativity.

Then, suddenly, attempts to construct a logical edifice for mathematics, the language of physics, fell in ruin.

Interpretations of reality in the light of these theories has been steeped in contentious metaphysics to this day.

Technical terms not defined in the text, shown in *italics*, are defined in Appendix A.

Pre-quantum history

Discovery is always an accident, but scientific discovery seldom happens to those without desire to find truth and design purposeful experiments to reveal it. The following is an account, for the benefit of non-technical readers, of background events that led to the quantum mechanics. Others familiar with this history may wish to skip to the next sub-chapter.

Two roads came to a confluence with a third in 1913. The first began when Isaac Newton held a glass prism in a shaft of sunlight coming through a hole in a blind. The prism cast a band of light on the wall ranging across the spectrum of color violet, blue, green, yellow, orange, red. Another inverted prism reassembled the colors back into white light. Newton's conjecture of what was happening was wrong.

The next milestone was an experiment by Young about 1800, described in more detail later, that proved conclusively that light propagated as a wave. A light beam split in two, recombined in a series of light and dark fringes where the light waves alternatively canceled and reinforced. The spacing between the fringes was twice as great for red light than blue showing the red wavelength was twice that of blue. A refinement of this experiment was the diffraction grating that enabled the wavelength of light to measured with very great accuracy. It falls in the 1 to $^1/_{10}$ micron ballpark from the infrared to the ultra violet, with visible light in between. A micron is $^1/_{1000}$ millimeter. Light is now known to be electromagnetic waves which include everything from radio waves to X rays and gamma rays.

The next milestone on this road, about 1815: Fraunhoffer repeated Newton's experiment but in a way that separated colored light into each exact wavelength. Light from a narrow slit was rendered parallel with a lens so that all light of a given color was refracted through exactly the same angle by a prism, a different angle for each color. Light emerging from the prism was imaged with another lens onto a screen where the slit was imaged each color separately. Fraunhoffer was no doubt disappointed to find that incandescent light revealed nothing more than Newton had seen. The payoff was sunlight, which also gave Newton's continuous spectrum of light from violet to red, but thousands of wavelengths were missing, appearing as dark images of the slit, called lines. Having no clue why these lines were missing, he classified them as A, B, C, D . . in descending order of prominence. Search for *Fraunhoffer spectrum* with your web browser for pictures. It was soon discovered that subsets of these lines could be reproduced in the laboratory both as bright emission lines and dark absorption lines caused by each chemical element, showing that the sun is made of the same stuff as we are. Well, there was this prominent set of lines due to no known element, so it was named after the sun. Then it, helium, was found coming out of a hole in the ground in Texas.

Each chemical element had a finger print of exact wavelengths of emitted or absorbed light, with little rhyme or known reason, with the exception of hydrogen. Johannes Rydberg in the 1880s found the empirical formula:

$$1/\lambda = R.(1/n^2 - 1/m^2)$$

represented the spectral line wavelengths λ of hydrogen, where m is a lager integer than n. R was whatever constant Rydberg needed to make it work.

We are now at the 1913 point of joining of the two roads, one described above, the other described next.

Thomas Edison theorized that the blackening of the inside of electric incandescent lamp bulbs was caused by electricity escaping from the filament and sticking to the glass. To test this theory, he made a bulb with a plate near the filament having its own wire coming out of the bulb. When he turned the bulb on, a feeble negative current flowed from the filament to the plate. He was right about electricity, but wrong abut the blackening which was caused by evaporation of the filament material. He was an inventor, using science to support invention of useful products. Seeing no practical utility in this discovery, he moved on to other things, but he was methodical and documented everything he did.

In the 1880s, John Ambrose Fleming with a new PhD, admiring Edison's use of electricity in lighting, joined Edison's company. Having access to Edison's notes, Fleming repeated Edison's experiment described above. Fleming put a battery in the plate circuit and found that when the plate was electrically negative, no current flowed at all. However, when plate, also called the anode, was positive, the current from the filament, also called the cathode, was a thousand times greater than Edison observed. Fleming's valve was enormously useful to Marconi because it was more reliable and a better valve than the crystals Marconi had been using in radio receivers. In 1916, Lee de Forrest added a grid to Fleming's valve, making it an amplifier. It was named the **Audion** by de Forrest, but British kept calling if a *valve* and Americans a *vacuum tube* because de Forest used glass tubing from a lab supply to build it. With the addition of more grids for various purposes, the audion launched the electronics revolution. It was the mainstay of electronics until well after World War II. By 1930 every home had a radio typically with five or ten audions. Now, replaced by solid state transistors, the audion has only niche applications with one big exception: The magnetron in microwave ovens and radar. It is a radial audion with the grid replaced by a magnetic field.

Fleming valves and audions, being common place devices early in the history of electronics, taught a valuable lesson in physics to anyone playing with them: If the anode voltage was high and you turned off the room lights, you could see a blue glow on the glass bulb where electricity, missing the anode edges or passing through a hole in it, struck the glass. If you put a magnet near the glow, it squirmed.

In 1897, J.J. Thompson knowing about the blue glow effect, did some real physics: He made the cathode ray tube: An elongated Fleming valve with a hole in the anode, made electricity into a narrow beam in a hard vacuum. Where it struck a glass screen at the end of the tube, it made a spot of light. Thompson could deflect the spot with magnetic and electrical fields showing the beam could be modeled as particles with mass and electric charge. Each particle had the same electric charge to mass ratio which his careful crafting of the apparatus allowed to be measured accurately. This represented the discovery of the electron. One Noble prize to JJ, please.

Robert A. Millikan watching a microscopically small oil droplet falling, could arrest the fall with an upward electrical field if the droplet was electrically changed. The droplet gained or lost charge in increments Millikan could measure, which he interpreted correctly as the electric charge of one electron. Combined with Thompson's measurement, the mass of the electron was known. Following established convention, the intrinsic electric change of the electron is negative. Give Bob a Noble too, please.

Ernest Rutherford showed that atoms, being electrically neutral, had particles with both positive and negative charge and mostly empty space. Stripped of an electron, an atom could be accelerated into a beam and deflected by magnetic fields, showing that the hydrogen atom is 2000 times heavier than an electron. Its nucleus is now called a proton. The same games played with other atoms corroborated the knowledge of the chemists measurements of relative atomic weight of the chemical elements, with the additional discovery that each element could exist as isotopes with different weights, but all with the same chemical properties. The two roads described above meet here, where Niels Bohr models the hydrogen atom. They join the quantum mechanics history road, described next. Nobles to Ernie and Niels also, please.

A brief history of the quantum mechanics

Technically savvy readers please go to the next paragraph. The rest of this paragraph is dedicated to preventing befuddlement by exposure to terms like geometry, kinematics, dynamics and mechanics. Geometry is what you need to know to build a pyramid or a chair: Spatial distances and angles of a fixed structure in space. Kinematics is what happens when geometry changes in time, as in cinema: light patterns on the screen change with time; moving pictures; pistons moving in the cylinders as the crankshaft turns. Dynamics is the kinematics of massive objects under the action of force. Add fuel to the cylinder: Potential energy. Convert energy to force: Needs spark-plug, runs good. Classical Mechanics: All of the above using Newton's laws and Euclid's geometry plus electromagnetic forces. These are the "Laws of Physical Reality". Relativity and Quantum mechanics: Wrenching reassessments of all of the above in which no obviousity can be taken for granted, including the demigod "Physical Reality".

The quantum mechanics developed over a three-decade period starting at the turn of the 20th century making its debut by 1930. The precipitating event was the discovery by Max Plank that light, which beyond doubt propagated as an electromagnetic wave, was emitted in lumps of energy, called quanta. He needed quanta to resolve absurd prediction, known as the ultra-violet catastrophe, made by what we now call classical physics, specifically by the marriage of Newtonian mechanics and Maxwellian electromagnetics. The latter was the foundation by James Maxwell in the 1800s, on four axioms, of all earlier theories of electricity and magnetism. Maxwell did for electromagnetism what Euclid, in antiquity, did for geometry. Euclid's geometry was based on five axioms, Maxwell's on four. In order to represent observed data, Planck showed that the each quantum conveyed *energy* E = h.f where f is the time frequency of vibration of the wave (the number of waves in unit time) and . means multiplication. h, now called Planck's constant, has a precise but very small value in conventional units of measure of angular momentum. Because the quantity $h/2\pi$ kept appearing in equations, it was replaced by the abbreviation ℏ. *Planck's constant* can mean either h or ℏ. (The constant 2π converts angle from cycles to radians). It is now is recognized as a fundamental physical constant like the velocity of light. Planck showed that the math worked: It removed the ultra-violet catastrophe that had flabbergasted every ranking physicist of the era. Astronomers, to understand the physics of stars, were helpless without an understanding of the interaction of matter and radiation. They could not fetch a shovel full of star and send it to the lab. However, Planck could not bring himself to believe the interpretation his discovery forced on him: That a wave delivered energy in lumps.

Albert Einstein, using straightforward reasoning, then showed that photo-electricity demanded that light also be absorbed as quanta. A quantum of light, now called a *photon* (named by others), joined the electron and proton in being recognized as elementary particles of nature. For this work Einstein got his only Nobel Prize. This put everyone into the metaphysical tizzy to which Planck objected because, beyond any doubt, light was known to propagate as a wave. You will find the term *fundamental* particle in older texts, but the proton and its mate the neutron are not now seen as fundamental because each they are composed of three quarks.

In 1913, walking in Newton's planets-orbiting-the-sun inverse square gravity law footsteps, Niels Bohr's mostly classical model of an electron going in an orbit around a much more massive proton under the inverse square electric force of attraction explained, in terms of measured physical constants, the wavelengths of light emitted or absorbed by atomic hydrogen. Bohr could calculate theoretically the Rydberg constant R, in the formula given earlier, using the data measured by Thompson and Millikan. The integer indexes m and n in Rydberg's empirical formula $(1/n^2 - 1/m^2)$ represented the energy levels of the electron. Bohr's model leaned heavily on Newtonian mechanics but added an ad hoc quantum assumption that allowed the model only to emit or absorb energy in the increments of the Rydberg formula: With no basis in known physics, the electron orbiting the nucleus could only have orbital angular momenta that were multiples of Planck's constant: The very constant Planck needed to quantize photon energies.

Transitioning from a higher to lower energy level, the electron emitted a photon. This would be seen as a bright spectral line. If the atom was exposed to light of all colors, the electron would absorb a photon when jumping to a higher energy orbit. Enough atoms doing this made the dark lines Fraunhoffer saw. From Planck's formula $E = h.f$, the energy difference E gave the frequency of vibration f and hence the wavelength λ of observed light of the hydrogen atom spectral lines. While Bohr only modeled hydrogen, clearly the same sort of thing was going on in all atoms of the chemical elements.

Bohr got a well deserved Noble prize for this piece of work. Where Heisenberg later succeeded, Bohr's model gave no accounting of the brightness of spectral lines.

Because of its wave/particle duality, the photon was an enigma. Everything else behaved properly, or so it seemed, until Louis de Broglie showed theoretically, using the Special Theory of relativity that all particles, even those of what we see as the material order, like electrons and assemblies of particles such as atoms and molecules, also propagate as waves. Subsequently this was demonstrated experimentally using wave *interferometers*: These waves have spatial frequencies (the number of waves in unit length) which, multiplied by Planck's constant, yield the particle *momentum*. This parallels Planck's $E = h.f$ except momentum replaces energy and spatial frequency replaces temporal frequency. De Broglie did not take us all the way to the quantum mechanics, but revealed why Bohr's ad hoc electron orbits were quantized: They were wave resonances and like a piano string will only resonate at discrete harmonics. A Nobel and PhD for Louis, please.

The reason photons and electrons have different wave formalism, is that they belong to one of two different particle families, Bosons and Fermions, named after Satyendra Nath Bose and Enrico Fermi. Bosons like photons have an intrinsic angular momentum, called spin, in multiples of h. Fermions like the electron, proton and neutron have spins in multiples of $h/2$. These two families obey different laws of the way *quantum amplitudes* (described later) are combined, giving these two particle families radically different properties. Fermions are the particles of the material order and Bosons the particles mitigating forces between them. Subsequently Dirac, Einstein and others showed that their spins appear naturally in relativistic versions of the quantum mechanics.

The funeral of Newtonian Mechanics

Towards the end of the quantum mechanics developmental period, Werner Heisenberg, one of the quantum mechanics principal founders, made the melancholy pronouncement of the death of Newtonian Mechanics, which had been the basis of dynamical physics for 250 years, with its astonishing successes in astronomy and engineering. The initial conditions for the solution of equations of motion could not be measured in a way that gave a unique solution to the equations. This was not a technical difficulty. It is fundamentally impossible to do so. Newtonian mechanics was at best only an approximation, albeit very accurate when applied to massive objects. In retrospect, we can see this as prognosis of certain death for the classical realm: Any rational basis for believing in the ultimate reality of the physical world died with it. However, a metaphysical basis has persisted through to the present. Consciousness, a mourner at the funeral, is immortal.

Newtonian mechanics has a simple basis in three laws: A force acting on a mass causes it to accelerate in the direction of the force by an amount proportional to the force and inversely proportion to the mass. This was Newton's second law. Also, every force is met by an equal and opposing force. This was Newton's third law which later was shown to be a consequence of the law of conservation of energy which appears to be a fundamental law of nature. Newton's first law is never used in setting up equations of motion – it is a special case of the second law and was needed for philosophical purposes: To contradict an error of Aristotle. Solving mechanical problems requires specifying the system's initial state and its state of motion at the same time: Heisenberg showed this to be impossible.

Equations involving rates of change are called *differential.* Acceleration is the rate of change of velocity and velocity the rate of change of position. For this reason Newton's second law mathematically is called a *second order* differential equation. Although it is easy write the equation, it is not always easy to find the equation describing the system state and motion that satisfies the differential equation. Telling you what the sun's force on Mars is doing to its acceleration does not directly tell you where to look for Mars in the sky, even if you know where Mars once was and how it was moving then.

In 1833 Hamilton showed that Newtonian mechanics could be reduced to two simple first order differential equations, called *canonical.* We can use any convenient *coordinates* $q_1, q_2,...$ to describe the system state, but once this choice is made it dictates that the system motion be described by specific *coordinates* called the *conjugate* momenta $p_1, p_2,...$ The constraint was that to be *conjugate* the product of a coordinate and its momentum be *angular momentum.* The distinction between "state" for the q and "state of motion" for the p is somewhat arbitrary. They are all coordinates, but come in *conjugate* pairs. The term *complementarity* was used by Bohr to describe the same thing quantum mechanically. Mathematically, complementary pairs are *Fourier transforms* of one another. They are described in a later chapter. Examples of such pairs are position and linear momentum, angle and angular momentum, and energy and time. The last example shows that thinking of "state" and "state of motion" can be fuzzy. This was a harbinger of things to come: In the quantum mechanics Planck's constant is a unit of angular momentum. Also one had to describe the system energy H in terms of the p and the q. H, called the Hamiltonian, is the sum of the *kinetic* and *potential energies* of the system.

Hamilton's differential equations of motion then become:

- time rate of change of p_i = negative rate of change of H with respect to q_i
- time rate of change of q_i = positive rate of change of H with respect to p_i

with a pair of these differential equations for each i = 1,2,3... Each value of i is called a *degree of freedom* of the system. A bead sliding on a wire has one degree of freedom, a hockey puck sliding on ice has 2, a bullet in space has 3, two bullets have 6. A curve ball has 6, because we not only have to account where it is in 3 dimensions, but also how it is spinning.

Math Footnote: The first of Hamilton's equations is related to Newtons second law. The time rate of change of the ith momentum on the left is the force equated to the ith component of the gradient of the potential or force on the right. The rates of change are partial derivatives (see Appendix C) so the kinetic energy does not contribute to this term. The second equation is the velocity in the q coordinate on the left equated to the rate of change of kinetic energy on the right with respect to the momentum p. Now the potential energy is a function of the q so the partial derivative with respect to p ignores it as a variable.

Solving these equations means finding how the p and q change with time which can be amazingly difficult given that Hamilton's equations are amazingly simple. For an actual system, say the earth-moon-sun system, we have to measure all the q_i and all the p_i **at the same time** to describe the system initial state. **Heisenberg showed that doing this was quantum mechanically impossible for complementary pairs.** In the quantum mechanics, the position q and its conjugate momentum p are given by operations Q and P. It is generally true that the order of operations A and B is irrelevant i.e:

AB = BA or, what is called the *commutator*: AB − BA = 0

```
(2 times 3 = 6) minus (3 times 2 = 6) = 2.3 − 3.2 = 0
(2 plus  3 = 5) minus (3 plus  2 = 5) = 2+3 − 3+2 = 0
```

So Hamilton's version of Newtonian mechanics would tacitly assume in effect that PQ − QP = 0. This is described by saying the operators P and Q *commute*. Heisenberg showed this not to be true, that PQ − QP = h = Planck's constant when p and q are complementary pairs. This equation is explained in more detail below. To be sure, h is very small so taking it to be 0 is a very, very good approximation if the mechanical system we describe is a planet, steam locomotive, a grandfather clock, or even a pinhead. If it is an atom, or an electron, Newtonian mechanics largely fails even though Bohr's largely Newtonian model got correct hydrogen energy levels.

Heisenberg showed that his observed data, the light emitted by individual atoms, called atomic spectra, could be represented by a *linear algebra*, subsequently shown to be *matrices* operating on *vectors* in *Hilbert Space*. His work was described as *matrix mechanics*. Here Bohr's model was useless.

Erwin Schrödinger and Paul Dirac showed that Heisenberg's *matrix* formalism was equivalent to the solution of a *differential equation* that embodied an energy operator **H**, analogous to the Hamiltonian of classical physics. Observations or measurements were represented by the results of *linear operators* acting on the solution to the equation called the *wave-function*. Heisenberg's matrices are equivalent to these operators. As the next paragraph relates, the anatomy of the new theory was significantly different from all earlier science. *Functions* and *differential equations* are described in Appendix C and F.

The Quantum Mechanics departs from earlier science

Early in the history of the mature quantum mechanics, E.U. Condon & G.H. Shortley, in their 1935 tome *The Theory of Atomic Spectra*, described the relationship between classical physics and quantum mechanics. This was one of the first texts presenting the new physics to scientists in the field. *In classical physics the laws are mathematical relationships between observed pointer-readings, as has been true in all earlier exact science. Quantum mechanics does not do this.* They did not give a précis of what quantum mechanics did, but with the benefit of hindsight we can say that apparatus and operations appear as operators which operate on the state of the system descriptor in a way that blurs the distinction between the observed system and the apparatus observing it. *We are dealing not merely with a new laws, but a new mathematical canvas on which to represent them. In this respect the quantum mechanics is a more far reaching departure from classical physics than were the theories of relativity.*

Brief history of the theories of relativity

In the same time frame that the union of Newtonian mechanics and Maxwellian electromagnetism, the cornerstones of classical physics, made catastrophically wrong predictions of the interaction of matter and the electromagnetic field, an experiment by Albert Michelson and Edward Morley to measure the velocity of the earth through space failed. The apparatus was simple and precise and Michelson's credentials were beyond question. He was later to win a Nobel prize in physics for other work. This failure in 1887 created a crisis in *kinematics* that was not resolved until 1905, when it was explained by Albert Einstein. During the next few years, Einstein slaughtered the three sacred cows of civil and scientific metrology: mass, length and time, seen as the absolute attributes of physical objects. The distinction between weight and mass was confused until Newton clarified it.

Einstein's first paper devastated kinematics which had been with us from antiquity, and a subsequent paper showed the relativity of mass, impacting Newtonian dynamics. His series of papers collectively is called the Special Theory of Relativity. It includes the relation of mass and energy: his famous equation $E = m.c^2$ concerning the equivalence of energy E and mass m, with the constant of proportionality being the square of the velocity of light. None of this had anything to do with quantum mechanics. But that was about to change.

Although the events that led to the Special Theory and the quantum mechanics started at about the same time in very different places, de_Broglie's early insight showed that the quantum mechanics cannot live without the Special Theory. The Special Theory was integrated into the non-relativistic quantum mechanics math by Paul Dirac.

In 1915 Einstein published the General Theory, displacing Newton's theory of gravity from fundamental physics. However, it has not been integrated into the quantum mechanics to date, notwithstanding a century of attempts to do so. Newton's gravitational force is transmitted at infinite speed, and in the General Theory at the speed of light.

What Einstein did

Around 1905 Albert Einstein did things that should have earned him three Nobel Prizes. He only got only one: Planck had shown that light was emitted in quanta, described earlier, but could not believe that they were particles. Einstein showed that absorption of light left no option but to understand that light was not only emitted but also absorbed as a particle. He reasoned: ultra-violet light can discharge electrically charged objects but red light cannot not no matter how intense, because the ultra-violet photon packs more energy than the red. That ultra-violet can discharge at any distance means that the photons deliver the same wallop no matter how far they traveled. This meant that each particle delivers the same Planck energy $E = h.f$ they departed with even though the number of arriving particles falls off with distance. A wave would fall off in intensity with distance and eventually be too weak to dislodge charge.

Energy and *momentum* are described in Appendix A. Because of their importance, some of their classical properties are cited here. The symbol . means multiplication, m is mass, v is velocity.

$$\text{energy} = \text{force} \cdot \text{distance}$$
$$\text{kinetic energy} = m \cdot v^2/2$$
$$\text{momentum} = \text{force} \cdot \text{time} = m \cdot v$$
$$\text{force} = \text{time rate of change of momentum}$$
$$\text{force} = \text{distance rate of change of potential energy}$$

The classical Hamiltonian is the sum of the kinetic energy and the potential energy. The kinetic energy is the energy of motion. Potential energy is what you store in a spring when you squeeze it; lift a heavy weight onto a high shelf; or is stored chemically in fuel before it is ignited. The conversion of potential to kinetic energy can be bad news to a mouse in a trap, or you when the weight falls on your head.

Then, in the same year, Einstein did something entirely different, for which he didn't get two Prizes he deserved, which led to something else in 1915 for which, in retrospect, he should have gotten a 4th Prize. His not getting these prizes came out of a prejudice of Nobel, who died in 1896, against the sort of physics Einstein did, seen as impractical or useless dabbling of a dilettante. After Noble's time a more enlightened view slowly emerged, but not quickly enough to give Einstein the prizes he deserved. This was the same shortsightedness of Edison who easily could have put a battery in the plate circuit; realized he could modulate the current with a grid between the filament an plate launching the electronics era, put Marconi out of business; measured electric beam deflection by electric and magnetic fields; got a prize Nobel would gladly have given him for discovering the electron instead of believing electricity was black stuff on the inside of lamp bulbs. These things were within the easy reach of Edison's ability. It is a solemn warning to young scientists: Do not do experiments to prove what you already believe is true. Do them to discover what really is true.

Einstein was motivated by a paradox he saw when he was 16: Maxwell's electromagnetic theory predicted electromagnetic waves moving with a velocity which was a constant for the substance in which the waves were traveling. This constant for various substances had been measured early in the 1800s, predating the electromagnetic wave idea. However, it was known to the experimenters that the constant they were measuring had the physical dimensions of velocity = distance/time. They were measuring the electric charge on a *capacitor* in both *electrostatic* and *magnetostatic* units of measure. The substance in question was between the plates of the capacitor. Various substances like glass, candle wax, wood, air were tested. The electrostatic force of attraction between the capacitor plates, measured by a precise balance, gave the charge in electrostatic units. The capacitor was then discharged through a *galvanometer* that had been magnetically calibrated. The deflection of a spot of light reflected from a mirror in the galvanometer gave the same charge in magnetic units. They were not numerically the same but differed by a factor related to the constant they were measuring. A vacuum is not a substance, but experimentally it gave a constant equal to the velocity of light. (The experimenters used room air, but a vacuum would have given essentially the same result). Although the speed of light was not accurately known at the time, the numerical coincidence did not escape their attention: a result which baffled them because the experiment had nothing to do with light. This coincidence remained a mystery for half a century.

It evidently occurred to Einstein that "moving through a vacuum" is meaningless - there is nothing there with respect to which one is moving – like asking: "How far away is nowhere?" Yet Maxwell's equations, which represented a huge body of historical evidence, predicted waves moving at the speed measured by the above experiment, but made no provision for observers moving with respect to one another. These waves were subsequently demonstrated experimentally by Hertz. Relative to what was the speed referred? "To space itself" most believed, soon to be shown to be metaphysical nonsense. Light moved relatively to the coordinate frame of each observer, to which the equations were referenced, at the constant given by this early experiment.

Nine years later, when Einstein was 25, he came up with one one of the simplest and most far reaching postulates of science: *Observers not being accelerated and all doing the same scientific experiment within their own reference frames, will all get exactly the same results, no matter what experiment they are all doing and no matter how they are moving relatively to one another.* He called this the postulate of relativity. A simple idea, but if true, it meant that light would travel in each laboratory at the speed dictated by the measured constant of the vacuum. The principle of relativity would require light wave from any lab to travel though any other lab always with the same velocity everyone had measured in the above experiment even if there was relative motion between the labs. This was inconsistent with known kinematics which allows velocities measured in different systems to be added like vectors. It also meant that the idea "moving through space" which Michelson and Morley tried to measure was meaningless: They were really measuring the velocity of their apparatus with respect to its own coordinate frame and got zero, obvious in the light of the above, but a metaphysically flabbergasting result to them.

Einstein stood on Newton's shoulders: Newton's laws were invariant in transformations between unaccelerated frames of reference in relative motion but this invariance did not call into question the laws of kinematics known since antiquity. Well before Maxwell, the laws of electromagnetics distinguished between what was "stationary" and what was "moving". Einstein's postulate made it universally true that what was observable depended only on the relative motion of the observer and what was being observed. In Galileo's time it was widely believed by the ignorant that "moving at high speed through space" was dangerous – an argument used against Copernicus – we would only be safe if the earth was at the center of the universe and not rotating.

The principle of relativity undermines mystery mongering not only about moving through space, but about rotation in it. It is meaningless to talk about something moving in a "straight line through space" or something "rotating in space". If two massive objects are in relative motion and there is no differential accelerating force acting on them and one of them is not moving in a straight line relative to the other, our frame of reference is defective by definition of Newton II which is quantum mechanically correct if the objects are massive. Such a frame is valid for physics only if the massive objects relative motion is a straight line in that frame. A massive object moving in a circle relative to the reference frame is centripetally accelerated. If there is no centripetal force causing this acceleration, the frame is not valid for mechanics because it violates Newton II. We can make a valid frame by firing three bullets with tail lights in the X, Y and Z directions, far away from massive objects. Our reference frame machine watches the bullets and keeps the axes pointing at them. Now we have a frame valid for physics experiments. The gods are laughing at us because they watch us all falling with enormous acceleration towards an infinitely distant, infinitely massive black hole and they see our straight lines as parabolic arcs. Our task is to find laws of physics that represent *our* experience: They do not have to account for things that amuse the gods.

Using the principle of relativity and the constancy of the speed of light, Einstein reformulated kinematics and got transformation equations for space and time measurements between two systems of measurement in relative motion. In the limit of low velocities, classical kinematics becomes an increasingly accurate approximation. The resulting transformations are symmetric between the two systems. So, if you and I are in relative motion, your meter stick looks shorter than mine to me, you will see mine as being shorter than yours. If I see your clock running slower than mine you will see my clock running slower than yours. This result is incomprehensible when the measurements are seen to describe absolute attributes of physical objects. This is illustrated by the twin paradox, described in detail later in Appendix G.

Einstein substituted these transformation equations into the equations of Maxwell's theory to see what they would look like in the other system and got, you guessed it: Maxwell's equations. In other words, the laws of physics look the same to all observers regardless of their relative unaccelerated motion. This simple idea, *the postulate of relativity*, will rank as a monumental event in all history. Michelson and Morley expected the observed speed of light to depend on one's velocity "through space". The form in which Einstein presented relativity violates the concept of physical reality deeply rooted in the intuition and common sense of scientists. The form in which it was presented by Newton called into question silly ideas of laymen. Newton stood on Galileo's shoulders, who understood the principles but not the mathematics. It was seen as heresy by politically powerful clerical authority intolerant of disobedience to dogma, as Galileo painfully discovered.

Lorenz Transformations

The transformation equations between systems in relative motion are called the Lorenz transformations. The failure of the Michelson Morley experiment caused a lot of shouting. Some said that the ether, a imagined medium that transports light, very light but very stiff because the speed of light is so high, we now know is very sticky too and glob of it is stuck to the earth with Michelson Morley apparatus stuck inside this sticky, stiff, light goo. Fitzgerald said no, no, the ether is very slippery and slithers between the atoms of the granite block the Michelson Morley apparatus is mounted on which changes the size of the block to just enough at all relative speeds to the ether to make the experiment fail.

The Leiden physicist Hendrik Lorenz looked past this hand waving, and understood that the Michelson Morley experiment would fail if the kinematic transformation equations between the moving and stationary systems left Maxwell's equations invariant, which takes an element of genius that by-guess-and-by-gosh-physics cannot replace. He did this before Einstein's work. Lorentz found that Fitzgerald's spacial transformation was not enough: Both space and time were needed to do the trick. They are not complicated, but the high school math need not be understood for present purposes. I have put them in Appendix B. Lorentz found these transformations by reverse engineering Maxwell's axiomatic equations, a fact unknown to Einstein at the time.

Not knowing what Lorenz had done, Einstein derived the same equations but from a fundamentally different basis. This quirk of history probably gave Einstein boost in acceptance: He was a 25 year old clerk, Lorenz a venerated professor at a major university. With Lorentz behind him, one could not give young Einstein the tempting brush off. Although Lorenz understood that this invariance would make the M&M experiment fail, he did not understand the basis of the equations he discovered, nor did ether believers, nor did Fitzgerald Einstein did.

When he published, Einstein added the *postulate of the constancy of the velocity of light in vacuum, c, with respect to all observers* to the *postulate of relativity.* Doing this was redundant, but he had nothing to loose in doing so, probably not wanting to economize on postulates by resting his entire case on the work of the early experimenters. They measured c as an electromagnetic constant of the vacuum and, in the light of the principle of relativity, would all measure the same constant regardless of their relative motion. In the light of Maxwell's equations, the same beam of light would be seen to be moving with the same constant speed even if observers were in relative motion along the direction of the beam.

Measurement of space and time

In the preamble to his first paper, understandable to a grade school child, Einstein pointed out two very simple truths about measurements in space and time that make them interdependent:

- Whenever we measure the **distance** between two points, we assume the observations of the correspondence between the points and the distance scale markings are made at the **same time** in the reference frame of relatively stationary scales and clocks.
- Whenever we measure the length of **time** between two events we assume that they happen at the **same place** also in the reference frame of relatively stationary scales and clocks.

This creates a quandary: In 1. we are measuring the time at two different spatially separated places, in contradiction to 2. We can put clocks at these two places, but how then do we know that they synchronize? The obvious answer is synchronize them when they are together, then transport them to the different places they are needed. In WWII movies we see soldiers all synchronizing their watches together before going separate ways on a mission that requires their activities to coordinate. Einstein had done his homework and realized that assuming the clocks would remain synchronized while they were being separately transported was dangerous. He devised this scheme: After putting the clocks where you need them, send a signal using light from clock A to clock B and echo it back to A. If it left A at 6 and got back at 8, we <u>define</u> its arrival time at B to be 7. This is a **definition** of remote time: We cannot claim to know independently that the signal got to B at 7, because we have no way of measuring it independently with clocks which have not yet been synchronized.

This has a space parallel. If we measure the distance from A to B by radar, we know when the signal left A but not when it arrived at B unless we already have a clock there that is synchronized. If we do not, we can time the signal from A to B and back to A which tells us the distance from A to B and back since we know the velocity of light. It is tempting to say that the distance from A to B is the same as the distance from B to A so we define the distance from A to B to be half the distance from A to B and back. But this is only a **definition** of distance, unless the clock at B is synchronized. One may say "why quibble about such obviousity" but if light travels at the same speed relative to all observers, and observers are in relative motion, obviousity goes up in smoke and Einstein knew this. He pointed out that if two systems of synchronized clocks are in relative motion, the clocks of one system will <u>not</u> appear to the other to be synchronized if the principle of relativity is true. This point is discussed in Appendix B and G. This flies in the face of classical kinematics and is the basis of the twin paradox (Appendix G).

Definition of the *same place*: We are not talking about Absolute Space which is metaphysics. If you have a system of metrology, scales and clocks relatively motionless with you. You look at the difference in time between two events at the same place in your system. I am streaking by with my system, these two events will not be at the same place in my system.

Einstein is often credited with the marriage of space and time. He deserves this credit to the extent of 1. and 2. above. But he missed this one: His thesis professor, Hermann Minkowski, made the amazing discovery that if time and space are considered a four dimensional space time manifold in which Pythagoras' theorem takes this simple form:

$$s^2 = +x^2 + y^2 + z^2 - t^2,$$

called the space like definition of s, but we can equally well use the time like definition:

$$s^2 = +t^2 - x^2 - y^2 - z^2;$$

either imply the Lorenz transformations, where x, y, z are space measurements and that of time t in units for which the speed of light is 1. It is amazing that Einstein missed this perception that space and time are geometrically woven together: The Lorenz transformations can be derived from the above equation of Pythagoras using high school algebra.

Einstein grabbed the ball from Minkowski and ran for a touchdown in 1915 with his General Theory of Relativity in which he made Minkowski's four dimensional space-time non-Euclidean in ways that explained gravity.

Non-Euclidean geometries were studied by mathematicians in the mid 1800s, but had no known scientific application at the time. They were seen as a paradox because mathematically they stood on as solid a foundation as Euclid's geometry in which parallel lines never cross. In Riemann's geometry they do cross which was seen to defy what was physically possible. Then Riemann discovered that a two dimensional Riemann geometry was mathematically equivalent to a 3 dimensional Euclidean geometry on the surface of a sphere, even though their parallel line axioms were in apparent conflict. Riemannian straight lines are *great circles*, straight because if you follow one it never veers to the right or left. If you make two slices through an orange the pie slice shaped sector edges will be parallel at the orange equator and cross at the top and bottom of the orange. But use a knife with a straight and not a wobbly blade or the cut lines in the orange skin will not be straight.

Pythagoras' theorem is more complicated when space-time is non-Euclidean and involves all of the mixed second order terms xy, xz, yt, etc. The properties of these spaces are described by 4x4 arrays called tensors which does not admit simple comprehension by ordinary people. Even the pros complain that they are "difficult". It took Einstein 10 years to master them. After he did, Einstein was elated for days when he discovered that his General Theory not only explained gravity, but to all observable accuracy correctly accounted for an anomaly in the motion of the planet Mercury: It had baffled astronomers because it was unexplained by Newtonian mechanics and gravitation. This was not as spectacular as the failure of the Michelson/Morley experiment because the error in Newtonian physics of Mercury's orbit was a very small effect known only to a couple of decimals of accuracy.

Coriolis force

Newtonian gravitation has been considered a fictitious force needed to correct the error caused by using incorrect Euclidean geometry instead of Einstein's correct four dimensional geometry. This has an historical precedent. Long range artillery shells did not follow the computed trajectory because the artilleryman's text book formula did not correct for the rotation of the earth. Coriolis pointed out that doing things correctly was mathematically messy. An easier approach was to introduce a fictitious force into the incorrect equations. Now called the Coriolis force, it made the shells veer off course as they were seen to do. A case where two wrongs do make a right. A major unsolved problem of physics is why the General Theory, in the century of its existence, has not been incorporated into the quantum mechanics, while the Special Theory was in the genetic code of the quantum mechanics so to speak and integrated by Dirac into Schrödinger equation by the time the quantum mechanics made its debut. Gravity, electrostatic and magnetic forces are members of the classical vector force field club. Gravity has defied inclusion in the quantum mechanics club: electromagnetic, weak and strong nuclear forces. This has raised the suggestion: Write off Newtonian gravitation as another fictitious Coriolis force. In the General Theory, a free body, like an astronaut floating in the space station experiences no force. So, if there is none one may ask, why try to quantize it? The same astronaut standing on the sidewalk does experience a force between their shoe soles and the concrete and is is not gravitational force but an electromagnetic force between fermions. One is left having to give a quantum mechanical answer to how mass causes flat Minkowskian space to warp to Riemannian.

The place of classical physics

Classical physics remains the preferred and simpler way to make predictions in its domain of applicability where the approximation that Planck's constant is zero is valid. There is a strong parallel with the approximation of architects taking plumb lines to be parallel, which is equivalent to a flat earth theory. Architects may make a living using the flat earth theory, but doing so does not require them to believe that the earth is flat. It would be a mistake not to use Newtonian gravity to compute orbits of astronomical bodies, even if a quantum theory of gravity were known.

Quantum Mechanics

The purpose of this chapter is not to teach the non-technical reader physics but to demystify the evolution of quantum mathematics.

Wave functions

The mathematical term *function,* as a relationship between numbers that can vary, is described in more detail in Appendix C.

The wave nature of the physical order lies at the foundation of the quantum mechanics. Light as electromagnetic waves were predicted theoretically by Maxwell in 1864 and demonstrated as waves later by Hertz. Subsequently these were the waves of radio. The wave did not go away when Planck and Einstein showed the need for light particles: photons. Their *energy* was locked to their wave frequency in time of the propagation wave.

The birth pains of the quantum mechanics were the conflicts between the apparently irreconcilable truths that light was both a wave and a particle. This quandary spread when de Broglie showed that all atomic particles, even complete atoms, also propagate as waves. In these cases their *momentum* was locked to the wave spatial frequency of the propagation wave. Even a bullet propagates as a wave but because of its enormous momentum the wavelength is so short as to be insensible. To all observable accuracy, we see it as a physical object at every instant of time with a definite place in space and a definite state of motion, which Heisenberg showed to be impossible.

Although Aristotle disliked empirical information, he used the curvature of the earth's shadow on the moon during lunar eclipses, that he had seen during a lifetime of observing them, as proof that the earth is accuracy a sphere. Yet many since have believed the earth is flat. Architects and builders to this day use a flat earth theory for most projects. This greatly simplifies building design and construction while making errors as large as the diameter of a gnat's eyelash. The building inspector never notices. Using a flat earth theory to design the Golden Gate bridge makes errors of only a few centimeters. The suspension cables expand and contract more than that from day to night. For most purposes the wave nature of the world can be ignored and replaced by a three dimensional space containing here and there substances made out of material points each with a definite position in space at every time, either stationary in that space or moving at a definite speed in a definite direction at every instant of time, or as science prefers to put it, at a definite velocity at every instant. Until the 20th century, this view was the metaphysical basis of matter. But it was believed to be physics: The physical world modeled itself as exact mathematical theory: the known laws of physics. It crossed no mind to see the laws of physics as mathematical approximation and the physical world an illusion created in its image until the late 19th century when physics ran into an empirical cliff, then fell into a theoretical abyss. Even to this day the physical world delusion limps on, bandaged with absurdity.

A quantum mechanical system state is described by a Schrödinger's wave function which correctly described hydrogen atoms. The earlier partly successful particle theory of Niels Bohr and various modifications of it by Arnold Sommerfeld, failed to represent all observed data, particularly brightness of light the atom emits only at specific colors. Instruments called spectrographs show each color as a line, called a spectral line. Look up *spectral line images* with your web browser for pictures that would not be done justice in gray-scale printing. Each color is a different wavelength. Light emitted by atoms appears as bright lines against a dark background. Light absorbed by atoms appears as dark lines across a continuous bright spectrum from violet to indigo, blue, green, yellow, orange to red. The spectral lines of hydrogen extend into the ultra-violet and infrared that the human eye cannot see, but can be seen with photographic plates and photocells.

In all the previous history of science, observations were pointer, scale or counter readings of some measured quantity such as time, fluid volume, electric current, etc. Outside the quantum mechanics this is still true of much of science. Classical scientific theory is the mathematical relations between these measured quantities. This is not how the quantum mechanics works. Schrödinger's wave function assigns a single number, called an *amplitude*, to every configuration of the particles that make up the mechanical system it describes, where a single configuration looks like a physical world model in space and time. The correct interpretation of the amplitude is due to Born, but has carried a metaphysical burden for nearly a century, notwithstanding the fact that the way to remove this burden has been known at date for the past sixty years.

What Heisenberg did

Werner Heisenberg reinvented the wheel of linear algebra to model atomic spectra: the emission and absorption of light at discrete wavelengths by atoms of the chemical element hydrogen. He was an assistant of Niels Bohr whose 1913 model of hydrogen, based on known physics, but requiring ad hoc additions, while correctly predicting the colors of spectral lines, provided no way of predicting their brightness. Heisenberg's great insight was to resist the temptation to invent a physical model to predict observable pointer readings, but rather to discover the mathematical relationships nature revealed to exist between observables. It takes the genius of people like Newton to play this game and win. It is our luck that Heisenberg's genius was at the right place and the right time. Like Armstrong's first step on the moon, this first step of Heisenberg, the birth of quantum theory, was a giant leap for mankind. Half a century later it was shown that no classical model can explain quantum reality. Although it took that long for the classical concept of reality to die, the lethal blow was struck by Heisenberg himself: The operators of complementary observables do not commute, explained below.

Heisenberg discovered that a summation over an index in each of the product of two quantities gave an array of quantities related to the brightness of spectral lines. Each summation over indexes of Heisenberg, described above, was understood later to be the *scalar product* of vectors in Hilbert Space: the amplitude for a system in one state to be observed in the other. More on this later. He left a paper describing this in the hands of Max Born for review and publication before leaving on vacation in Heligoland to relieve his allergies. Heligoland? Yes, hay fever.

Born recognized that Heisenberg was using operators called matrices by mathematicians: these emerge from the study of linear algebra. They are simply the coefficients of variables in linear equations written as a block, manipulated using certain mostly simple rules, described later. Heisenberg was not familiar with them, but Born was. Matrices as operators in linear algebra had been studied previously by mathematicians for at least a century. In the quantum mechanics, these operators are associated with the process of measurement. They are also used to transform from one system of representation to another in the same way that they are used to transform from one coordinate system to another in geometry. Linear algebra is described in Appendix E. Its basis is simpler than a lot of high school algebra, so matrices could be called grade school algebra - as long as you don't try to invert them. Then you need a PhD. Brain teasers in the Sunday paper often are equivalent of inverting 2x2 matrices which can be done using common sense while knowing nothing about matrices or linear algebra.

Born, with Jordan, seeded by Heisenberg, formulated the linear algebraic theory of the quantum mechanics, called matrix mechanics. Heligoland having cured his hay fever, Heisenberg rejoined Born and

Jordan in Göttingen. Born discovered that non-commuting operations that troubled Heisenberg were matrices called measurement operators P and Q of momentum and position: P operating on Q is not equal to Q operating on P expressed in the landmark equation of the quantum mechanics, often called the Heisenberg Uncertainty Principle:

$$PQ - QP = i.\hbar.I$$

i is the square root of -1 and $\hbar$ is Planck's constant. I is the unit matrix: It appears in the above equation to make it an equation in matrix algebra, not en equation in numerical algebra: It is just this array, which is just an array of numbers:

$$\begin{Vmatrix} 1 & 0 & 0 & 0 & \cdots \\ 0 & 1 & 0 & 0 & \cdots \\ 0 & 0 & 1 & 0 & \cdots \\ 0 & 0 & 0 & 1 & \cdots \\ \cdots & \cdots & \cdots & \cdots & \end{Vmatrix}$$

It represents a set of linear equations which operate on any vector to spit back the same vector: All the cross terms are 0 as you can see, in the linear equation each row of the matrix represents. The ‖ symbol is used to make a bracket around the matrix arrays using keyboard tools. The symbol | is used to make brackets for another purpose called a determinant. See Appendix E for more details.

Heisenberg's equation $PQ - QP = i.\hbar.I$ is truly remarkable as a law of physics. No wave-function appears in it. It is an equation in operators. It is important to see that when Heisenberg discovered it nearly a century ago, he was standing on the terra firma of physical reality underwritten by classical physics. The umbilical cord of fetus quantum mechanics was yet to be cut. To make certain predictions about the future of a classical system you have to know the coordinates q and the momenta p (there are as many of each that have to be known as the system has degrees of freedom) and they all have to be known at the same time. The only way to get them in the quantum mechanics to do experiments that perform the operations P and Q, but they cannot be done at the same time. Clever trick: Multiply the matrices P.Q = R. and do the operation R killing two birds with one stone. There. Or do Q.P = R, as you wish. Or, do one first then quickly do the other. Trouble is, if you do them in the order PQ, you will get a different result than if you did them in the order QP. So from the classical viewpoint you cannot make measurements that allow you to predict the single future deterministic history of the system. Ergo the Uncertainly Principle. **However, there is absolutely no uncertainty in the equation: PQ − QP = i.ℏ.I . It predicts uncertainty in only one place: The metaphysical big-rock-candy-mountain-land of Physical Reality.** The purpose of this book is to explain why there is no such place. Sorry, kids. But there is hope: Looking at candy can be sweeter than chewing it.

Heisenberg supplied the matrix mechanics link to physical observation, although he also had considerable mathematical insight. Born supplied the formal mathematics and Jordan the link to Hamilton's rendition of classical physics. Jordan saw what they were doing as the next step beyond Hamilton, described earlier. This was not borne out by subsequent history to the extent that the quantum mechanics was not the next step classical mechanics took; It was a radical departure from classical mechanics and all earlier science.

Only Heisenberg got the Nobel Prize in 1932 although Einstein also nominated Born and Jordan. Born is believed to have been excluded because of his association with Jordan who was a Nazi. The exclusion of Born upset Heisenberg, but Born eventually, well after WWII, got a Nobel for his statistical interpretation of *quantum amplitudes* which again sidestepped Jordan who died in 1980 with no Nobel. Without Heisenberg, matrix mechanics would not have happened or at best would have emerged slowly from the very different approach taken by Schrödinger.

See *Heisenberg's entryway to matrix mechanics* − Wikipedia

What Schrödinger did

Inspired by Planck and de Broglie and uncomfortable with the abstraction of Heisenberg's linear algebra, Erwin Schrödinger took a different approach modeling particle motion: He used de Broglie waves instead of particles as Bohr had done in 1913. In Bohr's particle model, the quantization of orbits was the ad hoc requirement that the electron orbits can only have classical angular momenta in integral multiples of Planck's constant, accounting for the observed hydrogen spectral lines using Planck's energy-wavelength formula. A decade after Bohr's model, de_Broglie explained why: electrons propagate as waves, so the Bohr orbits were wave resonances like the resonance of air in wind musical instruments or string in stringed instruments, which will only accommodate an integral number of waves: Ergo quantization. The significance of de Broglie's accomplishment was that he wrangled the wave nature of particles out of the Special Theory of relativity: He was not doing curve fitting to make Bohr come out right, even though it did so. De Broglie put it in a PhD thesis for which he not only got a degree, but also a Nobel Prize.

Schrödinger sought to model the hydrogen atom from scratch as a resonance phenomenon without, as Bohr had done, any concession to particle physics. Schrödinger's 1926 paper entitled: ***Quantization as an Eigenvalue Problem*** may be described modestly as a monumental event in the history of science, ranking with Newton's second law which, like Schrödinger's, cannot be derived from earlier concepts.

Schrödinger's famous time equation takes the form:

$$i \cdot \hbar \cdot (\text{time rate of change of } \psi) = H \psi$$

where ψ is the wave-function described above, **H** is a linear operator which embodies a description of the system, $\hbar$ is Planck's constant and i is the square root of -1. In the notation of Leibniz, the inventor of the infinitesimal calculus, we write this equation in this shorthand:

$$-i \cdot \hbar \cdot (\partial \Psi / \partial t) = H \Psi$$

where ∂ is not a number but is an operator which means an infinitesimal increment in what ever follows. So $\partial \psi / \partial t$ is the ratio of the increment in $\partial \psi$ to the increment in time ∂t that caused it: it is time rate of change of Ψ. See Appendix C for more detail. **H**, an operator, called the Hamiltonian after its classical cousin H which is a variable number.

ψ, the wave function, is a function of the coordinates of all particles making up the system and the time. If you have the equation for ψ and you plug in some numbers you concocted for the particle coordinates, the formula will spit out a number ψ. If the number is 0 the configuration you concocted is physically impossible. ψ is the amplitude, a complex number like x+iy, where i is $\sqrt{-1}$, to find the particles of the system at the specified locations at the specified time. ψ can be non-zero over extensive ranges of space and time. If you find the last sentence confusing, take heart: So did Schrödinger. An analogy would be the wind velocity of the atmosphere at a specified point in it at a specified time. If the wind had no vertical component, we could specify it exactly like a quantum mechanical amplitude with a complex number like v+i.w, where v would be, say, the east-west component of the wind and w the north south component.

H is not a number that multiplies anything, but a linear operator that describes the system energy in terms of operators which include the classical physics potential energy plus the system kinetic energy written in terms of momentum replaced by ***differential operators***. **H** may be complicated if the system is complicated. **H** operates on ψ which is a *function* of all the system particle coordinates and time: Why is the square root of -1 needed? Without **i** this would not be a wave equation.

Footnote:

H being a linear operator means that **H** operating on X + Y is the same as **H**X + **H**Y. **H**Y does not mean **H** multiplies Y but that the operation **H** is performed on the wave-function Y. Here, if we intend multiplication we will use a period . as the multiplication operator, like a.Y. We are representing operators in here **bold** type. Operators are not numbers, although a number followed by a . like 7. is the multiplication operator meaning multiply what follows by 7. This notation gives the wrong idea when applied to division and subtraction which provoked reversed polish notation ÷7, . 7, +7, −7 used on HP hand calculators. The operator idea goes back to grade school at least for arithmetic. The operator ∂/∂t means means the time rate of change of what follows. So ∂/∂t operating on Ψ means ∂Ψ/∂t which means the time rate of change of Ψ. Where confusion can cause one not to know which symbol is a variable (number, generally one that can change) and an operator, operator symbols may be written in a different type font.

Schrödinger showed the quantization of the electron energy levels as a natural solution to his equation, instead of having to add quantization as Bohr had to do as an ad hoc appendage to Newtonian mechanics. In the classical mechanics of musical instruments the harmonics called *eigenfunctions* and their musical pitches called *eigenvalues* appear naturally in the mechanics of string and wind instruments. A wave on a string, air resonance in a flute, correspond to de Broglie wave of the electron in hydrogen.

For the hydrogen atom, Schrödinger's equation can be resolved into two equations: One is time independent and gives the eigenfunctions corresponding to the Bohr orbits with eigenvalues equal to the Bohr energies. The time dependent solution describes what will happen if the system is started at time = 0 in an arbitrary linear superposition of eigenstates. This is usually done by starting the solution to the time dependent equation in one eigenstate and watching the ensuing evolution of linear superpositions of eigenstates as time goes on. This evolution is completely deterministic.

We are now at the frontier of physics and metaphysics.

The Amplitudes of Heisenberg and Schrödinger

Schrödinger knew he was on to something big because his theory correctly represented observations, but he was metaphysically confused, thinking at first that his wave equation computed the way electric charge of the electron was smeared out in space and time by the wave function.

Born realized that amplitudes did not represent any physical variable like electric charge. Amplitudes were not something you could measure with any kind of instrument, which is what measurements in classical science did. This was a major advance in our understanding reality. To make matters trickier, amplitudes were complex numbers of the form:

$$x.\sqrt{+1} + y.\sqrt{-1} \quad \text{usually written} \quad x + \mathbf{i}.y$$

Please see Appendix D for details about complex numbers. Born showed how the probability of observed events was related to amplitudes.

The quantum mechanics makes her debut

It had taken one-quarter century to get to this point in quantum history, but the quantum mechanics, after a decade's struggle by many brilliant people, was now blessed with an embarrassment of riches: Not just a theory, but two of them: Heisenberg's (not to speak of Born, Jordan, Pauli and Dirac) and Schrödinger's (not to speak of de Broglie). Within a year, Schrödinger had figured out that the two theories were equivalent in terms of their representation of measured data. The above events unfolded in 1925 and 1926. An interesting account of this equivalence by B.L. van der Waerden can be found online entitled *From Matrix Mechanics and Wave Mechanics to Unified Quantum Mechanics.* It includes an unpublished letter from Pauli to Jordan retrieved from a carbon copy.

What Born did

Apart from putting matrix mathematics into Heisenberg's quantum mechanics, for which Einstein believed Born deserved a Nobel Prize, Born pronounced that the amplitudes were related to the probability something would happen. Probabilities are ordinary positive *real numbers.* Born's formula to get a probability from an amplitude (x+i.y) was to multiply the amplitude by its own *complex conjugate*:

$$(x + \mathbf{i}.y).(x - \mathbf{i}.y) = x^2+y^2$$

or the square of the length of the vector (x, y) which is always positive real. It is also called the absolute square of a complex number. See Appendix D. It took a while, but Born was given a Nobel prize in 1954 ostensibly for this contribution to the quantum mechanics he made 30 years earlier, although he really contributed a lot more. If he had done nothing more, he still would have deserved the prize.

Born struck the final hammer blow that forged the quantum mechanics into a scientific theory that departed from all earlier science. A physical object is a configuration of particles with definite positions and momenta. If we thought of it in the wave-function context we might say ψ = 1 for that configuration and ψ = 0 for any other. We now know that such a thing is a fantasy. The wave-function in reality is non-zero over extensive regions of space and time when we use natural units in which ℏ=c=1.

Later we will reinterpret Born's probability as a *weight* pertaining to awareness. The word *probability* has a metaphysical interpretation which is unnecessary. You will commonly find the amplitude referred to as *probability amplitude*. The use of the adjective *probability* in this case, as will be shown later, is a concession to metaphysics, because there is nothing probabilistic about it. We prefer here to use the term *amplitude density* or just plain *amplitude*.

If the observed state is a continuum, such as the position of a particle in space, the probability of finding the system in some exact state is zero: We have to talk about a probability weight or density which is the probability of finding the system in a small volume of coordinate space which is the probability density times the small volume. We are assuming that function is smooth so that the probability density is pretty much constant everywhere in the small volume. We can multiply that constant by the volume to get the probability the system is in the configuration where the volume is located.

Whenever we talk of probability, we assume that the amplitude has been *normalized* to make the total probability to be found in all states to be 1 which amounts to saying that it is certain that the system will be found somewhere in the space of its variables. There is no big deal doing this: just multiply the amplitudes by whatever number makes the sum over probabilities come out to be 1. Jargon: Making the probability 1 is called *normalizing*.

Wave interference

Before the quantum mechanics, interferometers showed light to propagate as a wave. These instruments are described in more detail later. Light split along two paths when recombined and observed showed alternating bands of dark and bright, called fringes, where waves alternately reinforce and cancel.

This can be seen when two different wave trains meet on the shore of a pond.

Newton looked for light wave effects and pronounced against the wave theory of light when he could not see them. It did not occur to him to form the fringe image on his own retina where they are easy to see. In fact you can see them with apparatus no more complicated than two fingers held close to the eye – looking through a narrow crack between them at a point source of light like a distant streetlamp.

It is much more impressive if you look through two close slits as close together as you can make in aluminum foil with a mat knife. Look at the sun reflected in a distant convex mirror like a car bumper, or glass bottle, holding the foil very close to your eye.

Better apparatus with light of a single color would show this:

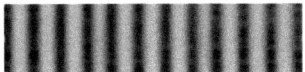

This was Young's experiment proving the wave nature of light. Before the quantum mechanics, after light was shown to be an electromagnetic wave, seen as a thing in the physical world, dark fringes were understood to be where the electric forces of two (or more) waves cancel, and bright fringes where they reinforce. This effect is called *interference* and an *interferometer* is a device designed to exhibit this effect. Above are descriptions of interferometers you can make yourself out of your fingers or a piece of foil.

Particles interfere with themselves. Paradox?

Classical electromagnetic theory represents accurately the behavior of large numbers of photons where one more or less would be insensible. The paradox of quantum mechanics <u>disappears</u> when you put lots of photons in a mental blender and blend them into a smooth creamy electromagnetic wave some of which goes through one slit and some of which goes through the other and wave interference is seen when the two waves recombine again. The paradox of quantum mechanics appears when you have only one photon: **It behaves in exactly the same way and interferes with <u>itself.</u>** Suppose we had made these discoveries in the reverse historical order. We would say "of course we see interference with lots of photons because each one interferes with itself". If one banana is yellow, a bunch of bananas also will be yellow. However, the metaphysical problem was understanding how in heck one photon can interfere with itself if it either has to goes this way or that way but not both. But if it goes both ways, it cannot be an indivisible particle. But it is! Paradox: There is something wrong with the way we are thinking: We are trying to imagine what we see as events played out in a physical world.

Quantum amplitudes

The quantum mechanics way of describing interference is to calculate the *amplitude* for a single photon to take <u>every possible path</u> through the interferometer separately to some point on the screen and add these amplitudes at that point together. These amplitudes can be large numbers everywhere on the screen, but where they cancel there will be a dark fringe. They can cancel because they are signed numbers.

If the amplitude to take one path is, say, $4 - 5.i$ and the amplitude to take the other is $-3.9 + 5.2.i$, adding these gives: $0.1 + 0.2.i$ For the above example the Born probability for the photon to be found in the dark fringe will be $0.1^2 + 0.2^2 = 0.05$. However in a bright fringe the amplitudes will reinforce giving a probability of $8^2 + 10^2 = 164$ which in this example, after the arrival of many photons is three thousand times as bright as the dark fringe. If we had normalized these numbers the sum of these probabilities would be 1, but the ratio of light to dark would be the same.

We have not described interferometers in detail. That will come later. However please note in the description above that the amplitude for <u>each</u> photon to reach each point on the screen is the amplitude for it to have gone through one slit <u>plus</u> the amplitude to go through the other.

These are signed numbers that can cancel in dark fringes or reinforce in bright fringes. There is nothing "probabilistic" about the fact that every photon uses both paths or all paths if there are more than two. This is the reason that the term *probability* amplitude to take each path is objectionable. It is certain that it takes all of them. The Born probability is a computed statistical variable that can be compared with the arrival rates of many photons at the screen or retina. If both agree we know our model of the interferometer is a good one. Later we will show that the computed amplitude, by summing the amplitudes for each path, is not probabilistic either.

Ignoring polarization and spin, the amplitude for a particle to go from one end of a vector distance **r** to the other is of the form:

$$\psi = [\cos(q) + i.\sin(q)]/r$$

This is a two dimensional vector in the Argand diagram (Appendix D footnote) and exhibits the wave nature of the amplitude. Its square is the Born probability and varies as the inverse square distance law. q can vary rapidly in space and time because $\hbar$, Planck's constant, is very small as the following rough dependence of q on space and time shows:

$$q = (\mathbf{p.r} - E.t)/\hbar = (\mathbf{k.r} - \omega.t)$$

p is the particle vector momentum, k the corresponding wave number vector and E its energy and ω the corresponding frequency in radians/second. If p is in the direction of r, k is the number of waves per unit length in the r direction. **p.r** means the *scalar or inner product* of **p** and **r**. It also shows that the amplitude can be non-zero over extensive regions of space and time. Scalar or inner product is defined in Appendix E. You can get by reading the following paragraph:

Given two lists of numbers, (a, b, c) and (x, y, z)

$$(a, b, c) \cdot (x, y, z) = a.x + b.y + c.z$$

is scalar product or inner product as it is sometimes called. Shown by the big dot between ( ) . ( ), commonly called vectors. The dot between a and x in a.x is ordinary multiplication of numbers. There should be no confusion using dot in both places because there is no way that ordinary multiplication of two lists of numbers has any meaning. There is a reason the word *product* is used: Scalar products obey all the algebraic rules of product algebra.

You are already very familiar with the scalar product of two vectors. You do it all the time helping mummy do her market shopping. Suppose you bought a apples, b bananas, and c carrots and the prices were x for an apple, y for a banana and z for a carrot, your market bill is the scalar product of these two lists. To pull rank on your baby brother, throw around jargon like "I'm helping mom by taking the inner product of two vectors". Remember: Linear algebra basically a list of scalar products, so don't let you quantum mechanist big sister pull rank on you.

Everything said above true of photons which are bosons is also true of electrons, protons and neutrons which are fermions, except for different rules of combining and calculating amplitudes. It is also true of assemblies of fermions like atoms and molecules. Long before we get to steam locomotives the quantum mechanics has blurred into classical physics so perfectly as to make classical appear as an exact science in the image of which some try to create God.

Equivalence of Schrödinger and Heisenberg

Non-mathematical readers: Read Appendix E before going on, being sure to understand what is meant by a scalar product. Assuming that you are familiar with the earlier appendices, Appendix F shows how to set up a classical differential equation for a stringed instrument and solve it.

What is important in a stringed musical instrument when the string is plucked, bowed or struck, is what it sounds like. What is important mathematically if the wave equation for the string is solved, is that it can be made to resonate in any one of its harmonics, six of which are shown below. The first is the fundamental or pitch note (bold), the second is the second harmonic (dot) which is an octave higher, the third harmonic (dash) is 3 times higher in frequency and so on forever. There are infinitely many mathematical harmonics. They are all anchored at the bridge and at the fret where the string cannot move. This restricts them all to be sine waves of the form: $y = a.\sin(n.x)$ where a is how loud the resonance is, x is the distance along the string which we will assume is π units long. Any other length is just as valid but complicates the math, so we use π. n is the harmonic number and is $n=1$ for the pitch note: the heavy black curve below.

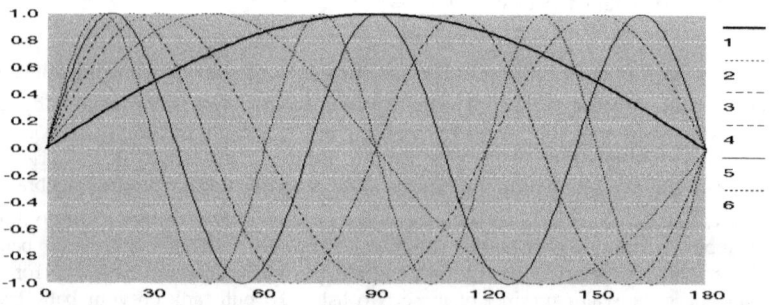

FRET   BRIDGE

The higher harmonics are 2 dot, 3 dash, 4 center, 5 solid, 6 dot, which is all the above figure shows. If the frequency of the pitch note (harmonic 1) is **f** cycles per second (256 for middle C), the pitch of any harmonic **n** is **n.f**. The horizontal axis in the above picture covers the range 0 - π but it is numbered in degrees so you can compare it with graphs of sine functions.

To go on reading, you have to agree to be force fed the following jargon. The harmonic is called an *eigenfunction* and the pitch is called an *eigenvalue*. Don't blame me, I am not German, but Hilbert and Courant were and they coined the term in 1924 to describe solutions to resonance equations that had been studied for a century. *Eigen* means *characteristic* or *proper*. The eigenfunctions are not always as simple as sine wave functions. Below are tabulated the six eigenfunctions graphed above at 12 points over the range of the half wave fundamental (pitch note).

The elementary differential equations of classical resonance and those of the quantum mechanics are *linear*. This means that if of the eigenfunctions e_1, e_2, e_3, e_4, ... each are solutions to the differential equation, and so is any sum of them and it is simply the sum of the solutions to each of them.

Multiplication is a **linear** operator:

$$3 \cdot (4 + 5) = 3 \cdot (9) = 27 = 3 \cdot 4 + 3 \cdot 5 = 12 + 15 = \mathbf{27}$$

However raising to a power is not a **linear** operation:

$$(4 + 5)^3 = \mathbf{721} \text{ is not} = 4^3 + 5^3 = 64 + 125 = \mathbf{189}$$

Since e_1, is a solution to our differential equation and

so is e_2, then because the equation is linear:

so is $e_1 + e_2$,

so is $e_1 + e_1 + e_2 + e_2 + e_2 = 2e_1 + 3e_2$,

so is $a.e_1 + b.e_2 + c.e_3 + d.e_4$ ..... where a, b, c, d ..... are any numbers, real or complex.

The above eigenfunctions for each note of musical instruments are exactly the same for flutes, guitars, clarinets, harpsichords, trumpets, pianos and tin whistles. The difference between the sound of these instruments is that each have a different set of coefficients a, b, c, d .....

Tabulation of y, the musical instrument string displacement for the above curves is shown below. x is point number, equally spaced, 0-π.

HARMONIC = EIGENFUNCTION

x	1	2	3	4	5	6
0	0.000000	0.000000	0.000000	0.000000	0.000000	0.000000
1	0.258819	0.500000	0.707107	0.866025	0.965926	1.000000
2	0.500000	0.866025	1.000000	0.866025	0.500000	-0.000000
3	0.707107	1.000000	0.707107	0.000000	-0.707107	-1.000000
4	0.866025	0.866025	-0.000000	-0.866025	-0.866025	0.000000
5	0.965926	0.500000	-0.707107	-0.866025	0.258819	1.000000
6	1.000000	0.000000	-1.000000	-0.000000	1.000000	0.000000
7	0.965926	-0.500000	-0.707107	0.866025	0.258819	-1.000000
8	0.866025	-0.866025	0.000000	0.866025	-0.866025	0.000000
9	0.707107	-1.000000	0.707107	0.000000	-0.707107	1.000000
10	0.500000	-0.866025	1.000000	-0.866025	0.500000	0.000000
11	0.258819	-0.500000	0.707107	-0.866025	0.965926	-1.000000
12	0.000000	-0.000000	0.000000	-0.000000	0.000000	-0.000000

<u>The following statement is of fundamental importance in the quantum mechanics and in metaphysical misinterpretations of it: All eigenfunctions are orthogonal to one another. This means that the scalar product of any one with any other is zero.</u> This is shown in the table below. The scalar product of any eigenfunction with itself cannot be zero because every term in the scalar product is a square and hence is positive. They can be normalized so that the scalar product of any eigenfunction with itself in 1, which I have done below, shown in bold type.

In the quantum mechanics there is a generally different set of eigenfunctions for each type of measurement.

SCALAR PRODUCTS OF EIGENFUNCTIONS $e_i \cdot e_j$
also called a Kronecker delta function of i and j written $\delta(i,j)$
for (eigenfunction i).(eigenfunction j)
i,j=1,2,3,4,5,6

i \ j	1	2	3	4	5	6
1	**1.000000**	0.000000	0.000000	-0.000000	-0.000000	0.000000
2	0.000000	**1.000000**	-0.000000	-0.000000	0.000000	-0.000000
3	0.000000	-0.000000	**1.000000**	0.000000	-0.000000	-0.000000
4	-0.000000	-0.000000	0.000000	**1.000000**	-0.000000	-0.000000
5	-0.000000	0.000000	-0.000000	-0.000000	**1.000000**	-0.000000
6	0.000000	-0.000000	-0.000000	-0.000000	-0.000000	**1.000000**

This table was computed from the table above at the full accuracy of my Intel processor, and gave ones and zeros to 16 decimals which I could not print in the space available.

$\delta(i,j) = 1$ if i=j. $\delta(i,j) = 0$ if i≠j.   That's orthogonality

It is true of the scalar product of **all** normalized eigenfunctions that are members of the same complete set.

Normalized simply means that each eigenfunction has been scaled to make the scalar product with itself come out to be 1.

<small>Footnote to math nerds: Strictly speaking I should have used the integrals of the product of the eigenfunctions as continuous functions but representing them as 12 component vectors is justified if there are no harmonics higher than 12/2, following a remarkable theorem of Nyquist: The 12 component vectors represent the function <u>exactly</u> because there are no harmonics higher than the 6th in this example.</small>

The above array interpreted as the coefficients of a set of linear equations is the Unit matrix: it transforms any vector into itself.

There is one final step to seeing the equivalence of Schrödinger and Heisenberg. It requires some understanding of the mathematical core of the quantum mechanics: **linear algebra**. This is described in more detail in Appendix E for those non-mathematicians who are interested.

Theorem: Any function ψ can be represented as a linear superposition of a complete set of normalized eigenfunctions e_1, e_2, e_3, e_4,

$$\psi = a.e_1 + b.e_2 + c.e_3 + d.e_4 + \ldots$$

To see why, suppose, given ψ, we want to know c the coefficient of e_3. Take the scalar product of both sides of the above equation with e_3. Except for the term $e_3 . e_3 = 1$ every term on the right disappears because e_3 is orthogonal to all the other eigenfunctions. This leaves

$$\psi . e_3 = c.e_3 . e_3 = c.1 = c$$

because: $e_3 . e_1 = 0$; $e_3 . e_2 = 0$; $e_3 . e_4 = 0$; $e_3 . e_5 = 0$; $e_3 . e_6 = 0$;

giving c where . means scalar product to get $\psi . e_3$. If ψ contained no eigenfunction higher than e_6 we could have tabulated it at 12 points and taken the scalar product of with e_3, which is $\psi . e_3$, as we did above.

Otherwise we have to get the area under the product of the two functions ψ and e_3 which is the calculus operation of integration. We can likewise get all the other coefficients **a, b, d,** .... The eigenfunctions must be a complete set. If we leave one out, we cannot make up for it by diddling the coefficients of the others because no linear combination of the rest them will synthesize the missing eigenfunction: They are orthogonal which also says they are linearly independent.

Hilbert Space

The vector (a,b,c,d,...) in the above example is a vector in a space called a Hilbert Space of dimensions e_1, e_2, e_3, e_4, ... so named by Hilbert's buddy John von Neumann. It probably should have been called von_Neumann Space but he was too modest to name it after himself. It seems like a stretch of imagination to name it after ordinary space, until one is sobered by the realization that ordinary 3 dimensional space is a special case of Hilbert Space.

We can now see the bridge between Schrödinger and Heisenberg:

When Schrödinger says ψ,

Heisenberg says (a,b,c,d,...).

Schrödinger's famous paper *Quantization as an Eigenvalue Problem* showed how the energy levels of hydrogen emerged from a pure wave model as eigenvalues as the harmonic pitches of a piano string emerge is the eigenvalues of the harmonics.

However, it was not immediately obvious to Schrödinger what Heisenberg's vectors were. After scratching his head for a year, Schrödinger realized that the two theories were equivalent.

Talking about these things is made easier using a notation invented by Dirac which is described next after some history.

What Dirac did

Heisenberg published his matrix mechanics in 1925 and later that year Dirac published his developed version of quantum mechanics. Schrödinger's 1926 equation did not account for the effects of the Special Theory of relativity. Dirac cast it into relativistic form in 1928 and showed the necessity of electron spin and predicted, as a natural outcome of the Special Theory of Relativity. His equation gave a rigorously correct description of the hydrogen spectrum, including the fine structure of H spectral lines. He pioneered quantum field theory and got the ball rolling on the quantization of gravity – a still unsolved problem – by showing how to describe quantum mechanics in non-Euclidean space. His work led to a theory of electron spin equivalent to that of Pauli. Uniting Relativity and Schrödinger's theory, instead of the messier mathematics one would have expected, led instead to an equation of great elegance and simplicity giving deep insight to the quantum mechanics of spin ½ particles where conservation of parity is valid. It revealed the existence of the anti-electron: the positron years before it was observed. Dirac took us more deeply into quantum reality, not to return to the familiar territory of classical physics, as Einstein may have wished:

Later recognized as autistic, the young Dirac was considered a sort of mad scientist by his peers. They invented a new unit of measurement of taciturn speech: the dirac = one word per year. At a public lecture, during the question period, an attendee said he could not understand a certain equation of the blackboard. After an embarrassing silence the moderator asked Dirac if he was going to answer the question. Dirac replied, in effect: "No, it wasn't a question. It was a statement". History does not tell us whether he had his tongue in his cheek. Bohr asked him if he ever got tongue-tied explaining things. Dirac answered that he did as he was taught in grade school: never to start saying anything, unless he knew how how was going to finish saying it.

His work has been ranked with that of Newton and Maxwell. Dirac shared the Nobel Prize with Schrödinger in 1933.

In 1930 Dirac published the first text on the quantum mechanics: ***The Principles of Quantum Mechanics.*** In a later edition he added his invention of his vector notation that facilitated the mathematical principles of the quantum mechanics. It embodies symbolic manipulation of Hilbert Space vectors in the quantum mechanics and their use in computing amplitudes. His notation has been the standard to this day. If a system is in state A and we ask for the amplitude for it to be measured in state B, we write this scalar or inner product as a bracket

<B|A> = bra ket = bra|ket

It represents one equation in linear algebra: The scalar product of B with A or the projection of A onto B, but with this twist: We can take this bra|ket apart and call |A> the ket vector and <B| the bra vector. However, if B = A we want the amplitude to be a real number 1, but the elements of a can be complex numbers. We get 1 only if the bra vector <B| is to be computed using the complex conjugate of B which is written B*. This means changing the sign of i everywhere it occurs in the elements of B. This is the definition of a bra vector, so we write <B| and not <B*|. Often one wants to transform the ket |A> to some other representation by multiplying it by a matrix, say **M.** Then we write this in the compact form <B|**M**|A>. This is why the bra|ket notation is read from right to left: We start with A and end up with B. The term representation is more precisely defined later. The outer product is written |A><B| is an operator which we need not discuss here.

Dirac pointed out that while there are a lot of parallels between vectors in geometric space and vectors in Hilbert Space, there are some marked differences. In ordinary space if you multiply a vector by -1 you get a vector pointing in exactly the opposite direction. So south = − north; west = − east. This is not true of Hilbert Space. You can multiply an eigenvector by any constant, any sign, even complex, and it still points in the same direction in Hilbert Space, at some eigenstate. If a wavefunction is equal to a linear supposition of eigenfunctions $y = a.e_1 + b.e_2 + c.e_3 + d.e_4 \ldots$ this equation is not altered if we multiply the right hand side by any constant, except 0. It still says the same thing as long as it does not change the ratio of a to b to c to d, and so on. If in the above equation e_2 points at bananas and b = 2, you have 2 bananas. If b = 7, e_2 still points at bananas. b = 0, says yes, we have no bananas, and I don't mean we have no oranges. -1 says I want a banana. A simple example of how Dirac's algebra is used is shown in the chapter Electron Spin towards the end of this book.

The rules of the quantum amplitude game

$$\psi = a.e_1 + b.e_2 + c.e_3 + d.e_4 + \ldots$$

is called the *state vector* of the system it describes. The numbers a, b, c, .. are called the *amplitudes* for each eigenstate they multiply. After doing that please <u>don't</u> say: "Now add the whole mess together to get ψ". Instead please say that *"it is a linear superposition of eigenstates"*. In case you forget what *linear superposition* means, it means: **add the whole mess together.** And please describe what you are doing as *linear algebra*, not mess addition.

Suppose you have some other state referred to the same eigenstates, say,

$$\phi = \alpha.e_1 + \beta.e_2 + \kappa.e_3 + \delta.e_4 + \ldots$$

and you want to know the amplitude to measure ψ in the state ϕ. Dirac says it is $<\phi|\psi> = a.\alpha* + b.\beta* + c.\kappa* + d.\delta* + \ldots$ where the * remind you that you have to use the complex conjugates of the amplitude they mark in the <bra| vector. If you do not do this, and because the amplitudes can be complex, you will not get 1 as the amplitude of a system in some state to be measured in the same state. Here 1 means it is certain to be.

The amplitude for a particle to follow a chain of linked paths is the <u>product</u> of the amplitudes for each link in the chain. Each link in the chain could be the path taken by a particle going through complicated apparatus of concatenated elements. The methods are understood for calculating the amplitude given drawings and specifications of the apparatus. If an event can happen a number of different ways, the amplitude for it to happen is the <u>sum</u> of the amplitudes to happen each way separately. Every path a particle could have taken must have its amplitude evaluated.   If there are two particles that do not interact and there is an amplitude for one to do this and the other to do that, the amplitude for both events to happen is the product of these two.   **The quantum math of compounding amplitudes could not be simpler. Except that the amplitudes are complex numbers, it is grade school arithmetic.**   But the kid has to be given the amplitudes. Computing them in the first place is a job for grad students.

There are further simple and very interesting rules for calculating the amplitude for events to happen that depend on whether the particles are bosons or fermions, which explain why these two families have very different properties.   The conclusions of the book do not require understanding these rules. If you are interested, you will find a good discussion in Feynman's chapter 4 of volume III of *Lectures on Physics* which can be read online.

The product rule for concatenated links means that the entire past history of the system compresses into a single complex number $x+iy$. We do not have to know anything else about the past. The math gets thicker when the number of paths is a continuum: infinitely many paths. To read this book you do not have to know how to add and multiply complex numbers, but here's how anyway:

```
(a+b.i) + (x+y.i) = (a+x) + (b+y).i, just another complex number.
(a+b.i) . (x+y.i) = (a.x) + (b.y)i² + (a.y +b.x).i
But i² = -1 because i is the square root of -1, so
                  = (a.x-b.y) + (a.y+b.x).i
just another complex number.
This is the same thinking taught in grade school to multiply 12 by
3 = (1.3).10 + (2.3).1 = 36
```

Eigenfunctions of an operator

In the quantum mechanics, what is observable is not directly pointer readings like the time on the face of a clock that the hand points at when some event is timed. Suppose some wave function ψ correctly describes the system and you wish to observe some measurable of the system. Associated with that measurable there is a linear operator, say **A**, that operates on ψ written just **A**ψ . (Wave functions don't have to be named with the Greek letter ψ , but the Schrödinger/Heisenberg gang started it and no one since has had the chutzpah to change. Ψ pronounced psi makes the **ps** sound as in ecli**ps**e or ecliψ)

If **A**ψ = $a.\psi$ where a is not an operator but a simple real multiplying constant, then ψ is an **eigenfunction** of the operator **A** and a is its **eigenvalue,** which would be an observed quantity in the quantum

57

mechanics. In other words, **A** operating ψ does not change it to another vector but only gives back the same vector ψ multiplied by a constant a. Because a represents a measurement which is always a real number, the observation represented by **A**, requires **A** to have a property called **Hermitian** named after mathematician Charles Hermite. Hermitian operators as matrices have a skew symmetry about the main diagonal. This is not complicated but need not be understood for our purposes. You can look it up online.

In the quantum mechanics the process of observation is described by an *Hermitian* operator operating on the wave-function. An operator does not change the observed state of the system if the wave function is an eigenfunction of the operator. If the state vector is an eigenstate the measurement operator is "deterministic" which means that repeating the observation will always give the same result. We are walking on the edge of a metaphysical precipice: Words in quotation marks warn you you are close to the edge. To be more exact, if a system is evolving in time in Schrödinger's equation, i.e. the wave-function is changing in time, if you make an observation at some time T say and find it in an eigenstate of the observation, if you make another observation quickly before the system has time to change, it will be found in the same eigenstate. In some cases there is no change in time, like a photon comes out of a Polaroid sunglasses vertically polarized, it will go through every other vertical polarizer for eons to come. In quantum mechanics, two observables can be known "deterministically" only if none of the measurements "changes the state of the system". Two observables can be known "deterministically" only if they have a common set of eigenfunctions: the outcome of the measurements is independent of the order in which they are conducted. This is described by saying the operators *commute*. If two operators commute they have the same eigenfunctions.

If the above paragraph seems to you as gobbledygook, this amazing fact pointed out by the great mathematician Euler (1707-1783) may help: There is a tennis ball on the table. You pick it up, play catch with it, then play some tennis and put it back on the table in exactly the same spot: <u>There is some diameter of the ball that will return to its exact same position.</u> Everything you did is equivalent to some simple rotation of the ball about this diameter. Call this diameter line vector **e** fixed in the ball. Call the total thing you did to the ball operation **A**, then **e** is an eigenvector of the operation **A**. If catch and tennis were operations **C** and **T** in that order then the operator **A** = **C.T** . The **.** here means matrix multiplication: Every element in the ith row and j^{th} column of **A** is the scalar product of the ith row of **C** with the jth row of **T**. This in not true generally if you commute the order of the operations like **T.C** if they have different eigenvectors. You can convince yourself of this using a matchbox instead of a ball. If **P** puts the matchbox back on the table with some line fixed in the box unchanged, and **Q** puts the matchbox back on the table with the same line unchanged, then you can commute the order of **P** and **Q**, because they rotate the matchbox about the same line which is an eigenvector of both **P** and **Q**. If **P** rotates 10 degrees and **Q** another 20 degrees the total is **PQ** = 30 = 10+20. Or you can change the order in which you did **P** and **Q** and get the same result **QP** = 30 = 20+10. But: Try 90 degree rotations say about a line through the face of the box for **P** then by 90 degrees through the striking edges for **Q**, the matchbox will end up in different positions if you commute **P** and **Q**. The operators **P** and **Q** now do not commute because they have different eigenvectors. In this example the matrices are 3 x 3 arrays, that is they represent three linear equations in relating 3 element vectors.

Historically, the mathematician Hilbert recognized that eigenfunctions can be treated as vectors which have multiple dimensions of a space now called Hilbert space: They are orthogonal which means: no eigenfunction can be represented by any *linear superposition* of the others. This is equivalent to the math jargon "each eigenfunction is *linearly independent* of all the others or *orthogonal* to the others". The quantum mechanical system state can thus also be represented by a vector in Hilbert space. A vector in Hilbert space is a list of numbers that are the multipliers of the eigenstates. If it describes a system state it is called the *state vector*.

Although Hilbert delivered lectures on *The Foundations of Physics* and wrote a book with Richard Courant on the mathematical methods of physics, he is not credited with being a major contributor to the development of the quantum mechanics, but he certainly knew what was going on – it used his math and von Neumann was his assistant. While working on the General Theory of Relativity, Einstein got some math help from Hilbert with the result, in seeing what Einstein was up to, he tried to scoop Einstein: In the end, however, Hilbert gave Einstein full credit for The General Theory.

Generally Hilbert spaces may have any number of dimensions, even infinitely many, or only two. A piano string has infinitely many harmonics, mathematically. An eigenvector is thus a unit vector pointing along some "dimension" in the Hilbert Space. The analogy in ordinary space may be east, north and up. Orthogonality in ordinary terms for example means no amount of going north and east makes any progress going up. Nor does any north and up travel make any progress east, and so on. Jargon: North, Up and East are linearly independent directions. They are *orthogonal*. Saying that you can go anywhere means that the eigenvectors are a complete set. I do not know of a simple proof that eigenfunctions form a complete set. But one can easily see the trouble you would be in if it were not. A helicopter that could go anywhere except up and down would not be of much use as a helicopter. But you could use it for fishing on the lake. At least you would know it could never sink. The Wikipedia article *Eigenvalues and eigenvectors* has a good active display of a simple Hermitian linear operator with two orthogonal eigenfunctions with distinct eigenvalues transforming a vector space in a way that does not change the direction of the eigenvectors but only scales them by their eigenvalues. Space down to the third diagram. This article describes how the eigenvectors and eigenvalues of a matrix operator are determined.

Observations in the quantum mechanics

You parallel park your car and run into a store. Returning to the car you discover a new yellow dent and some broken headlight glass, but the culprit split the scene without leaving a note. The police identify the make and model number of his car by reassembling the broken pieces. Knowing the color they can look at a list of registered cars in your locality to narrow down the search. The quantum story of observation is similar. You and your observing apparatus is one system and the thing you are observing is the other. Initially each is described by a wave function which gives the amplitudes for each system to be in every possible state. Before the interaction that constitutes the observation, the total wave function is the product of each component. During the interaction each changes and after the interaction the observer is left in some state: If the experiment is well designed, that state tells you something about the state of the observed system.

In the laboratory one usually gets observational data not from one observation but a large number of similar measurements: The results tabulated in what is called a density matrix. It does not describe a pure state but is a statistical device that represents a superposition of pure states. It is not of interest in what is discussed in this book.

A pure state of the observed system is the wave-function ψ of the observed system, expressed as a linear superposition of the eigenstates of the observation operator. Call this **A**, using bold type for operators. Different types of observation have different operators, say **B**, **C** and so on. If ψ has only 2 degrees of freedom these operators can be represented as 2 x 2 matrix arrays which have 2 eigenvectors with measurements represented by 2 eigenvalues. Depending on the complexity of the observed system the number 2 above may be any other integer and even be infinite. For example, a hydrogen atom has infinitely many theoretical energy levels but such an atom would be as big as the universe. Atoms are always being jostled by others nearby which would knock the valence electron loose, preventing the infinitude of higher levels from being occupied in any atom able to keep all its electrons.

Any mathematical function with the same number of degrees of freedom as the observation operator can be represented by a linear superposition of the eigenfunctions of the operator, discussed earlier. So the wave function ψ of the observed system can also be represented by a linear superposition of the eigenfunctions of the operator **A**. Call these $e_1, e_2, e_3, e_4, \ldots$ So:

$$\psi = a.e_1 + b.e_2 + c.e_3 + d.e_4 \ldots$$

and the state vector of the system is $\psi = (a,b,c,d,\ldots)$.

It may be possible to prepare a system in some definite eigenstate. For example an election can be prepared with its spin aligned to the field of a magnet in the same way a compass needle aligns to the magnetic field of the earth. This is true because the electron behaves like a bar magnet aligned with its spin axis. Quantum mechanically we cannot do an experiment to find what direction we aligned the electron, but we an ask the electron to tell us whether its spin is aligned in a particular direction of the observing apparatus. If our experiment asks the electron whether its alignment is in the direction we prepared it, it always says yes. More generally, if some eigenstate, say e_2, of the apparatus corresponds to the prepared state, the state is we determine:

$$\psi = 0.e_1 + 1e_2 + 0.e_3 + 0.e_4 \ldots$$

So far we have been talking physics, or in vector notation $\psi = (0,1,0,0,..)$

If we ask if the electron spin is in some other direction, it sometimes says yes and sometimes says no. More generally, if the state of the observed system is not just one of the eigenstates of the observation, but $\psi = a.e_1 + b.e_2 + c.e_3 + d.e_4 \ldots$ where the coefficients a, b, c, d.. of this linear superposition of eigenstates are not zero, sometimes we will observe it in e_1, sometimes e_2, and so on. The best we can do is to make a lot of observations and confirm that the probability of finding the observed system in some eigenstate of the observation is the absolute square of its amplitude or, what amounts to the same thing, the product of its amplitude, say a, with its complex conjugate a*, because a,b,c.. generally are complex numbers. If we make many measures, in this case the probability of finding the system in state e_1, will be aa*, and so on.

Understanding this was Born's insight. If the experiment shows the observed system to be in state e_3, and we remeasure the system again before it has time to evolve or interact with anything but our apparatus, it will still be found in state e_3.

We enter the realm of metaphysics when we try to understand why. The rest of this book explores the answers to the question.

A discussion of the issues raised by observation in the quantum mechanics may be found in the 2005 Bachelor of Arts thesis *The Role of the Observer in Interpretations of Quantum Mechanics* by Paul Sonenthal of Williams College.

In classical mechanics we can uphold ourselves as impartial observers of a world external to ourselves that we can see objectively. OK, we admit, we are made of the same stuff it is made of, but always we can make this detachment from it. Is this really true? No, to the extent that an approximation is never the truth. To the extent that doing so is a very accurate approximation to observed data we can construct a theory. If we create reality in the image of such a theory, it may paint a picture that is totally wrong: Example: The flat earth theory, an excellent theory for building contractors, but useless for geodesy

Tycho Brahe made extensive observations of Mars. Kepler used them to derive the laws planetary motion. Kepler did not have to worry about the extent to which Brahe's looking at Mars changed its orbit. Three centuries later Heisenberg showed that the observer and the observed generally cannot detach as we can detach Tycho gaze from Mars' orbit.

How can my looking at a little electron make such a big change in me. Mars moves in the same orbit whether Tycho Brahe looks at it or not. He saw some photons. Please contemplate the following quantum mechanical profundity: Tycho's looking at Mars didn't change Mars, but some photons from Mars did make an enormous change in awareness: To Tycho, Kepler, Newton, science and then the rest of history, finally even the Pope and the Catholic Church. By the time awareness reaches higher brain function the Planck's constant sized shenanigans of a particle have been amplified a trillion fold and there is no forgetting what happened.

More realistically, the laws of physics describe the relationship between you and the electron in the spirit of the principle of relativity, without assigning absolutism to either the observer or the observed and without distinguishing you from it unless you are no longer interacting.

Summary: In quantum mechanics, if the apparatus is set up to observe each eigenstate of the apparatus observing the system, **only one of its eigenstates will be observed of those included in the predicted linear superposition with each observation.** If the amplitudes of several eigenstates are calculated to be non-zero at the time of observation, and exactly the same experiment is repeated, it is likely to be found in a different eigenstate at the same corresponding time. We are using the word *found* as synonym of *observed* – something we can write down and be aware of.

This gives rise to the metaphysical conclusion:

1. *The eigenstate observed was selected by a random process which also kills the others that were not observed.*

The above statement makes the metaphysical assumption that one is always aware of everything that is real.

Everett showed that the above interpretation is metaphysics, not physics. It should read (my words)as follows:

2. *In quantum mechanics, if the apparatus is set up to observe each eigenstate of the observing operator, there will be a separate state of awareness for each eigenstate.* One could leave it at that, but if one does, those who look askance at the discussion of reality in physics, will proclaim: *I will be aware of only one eigenstate of those included in the predicted linear superposition with each observation and anything I am not aware of is not real.* This is the metaphysical trap in which physics has been caught, when sixty years ago, Everett did leave it at that.

The rest of this sub-chapter is a preview of where we are going. It uses the words *conscious, aware, real, exist* which are only given concrete definitions later. Vagueness will result if you define these differently than I. Defining them first creates a chicken and egg problem. Here is a little chicken to lay a big enough egg to hatch the real chicken later.

If the amplitudes of several eigenvectors are non-zero at the time of observation all of the eigenstates will give rise to equally real states of consciousness but they will not bind in common awareness. If exactly the same experiment is repeated from scratch, exactly the same superposition of conscious states occur with every repetition. There is no difference in reality between one repetition of the experiment and any other that starts in the same state. Exactly the same thing happens in reality with each such repetition. There is no random selection of different paths with each repetition of the experiment.

The difference between the above interpretations is:

The first <u>assumes</u> that there in only one path through reality from the past to the future: This is the metaphysics of classical mechanics upheld by Laplace. The Copenhagen Interpretation preserves it by relegating to non-reality all paths that do not bind in awareness with the one experienced. But, unlike Laplace where only one path could be real without violating physical law, now any path with a non-zero amplitude is permitted by physical law. Forearmed with the metaphysical prejudice "I am aware of all that is real", we cannot in this interpretation predict what will happen next, so randomness must rule reality. Because any non-zero amplitude path could have been real, our becoming aware of the one we experience must have "collapsed" all the others. This is irksome if they constitute awareness at remote places which require superluminal extermination. This is the metaphysical road Heisenberg took.

The second recognizes that each eigenstate the observation results in equally real but different history in consciousness. If the eigenstates are a continuum, the histories in consciousness are a continuum although each is discrete. We cannot observe consciousness because it does not exist. But experience of it is absolutely real. The quantum mechanics for the first time in the history of science gives us a beautiful view of the structure of conscious experience of awareness: Each state of awareness is caused by the projection of the state vector of the observed system onto each eigenstate of the observation. This model restores strict causality abandoned in 1. above. Because the eigenvectors are orthogonal, the states of consciousness of each does not bind in awareness with any other. For example one can be consciously aware that:

1 the electron spin observed is up and not down

2 the electron spin observed is down and not up

<u>These two realities cannot, without absurdity, happen in the same state of awareness, but they can be equally real in consciousness.</u> This is unbelievable to the Physical World meta-physicist because he believes it <u>is</u> reality and cannot be in inconsistent states. He is not denying the reality (or something) of consciousness, but follows in von Neumann's footsteps by tucking consciousness into the physical world, which he believes is is the real reality.

It is meaningless to talk about causality "in reality" but it is meaningful to talk about it in a theory that represents experience. We can consciously experience a correlation: A spark plug ignites fuel vapor in a cylinder. This happens a million times a second around the world 60*60*24*365*100 times a century. The philosopher Hume told us that does not prove causality. But we <u>can</u> put causality into our spark plug theory which claims it <u>always</u> works. If the spark plug does not ignite the fuel vapor just once, that instance disproves the theory as Hume reminds us, but it only fails once a century, it as still a very good theory, even if it is wrong.

The only role that the human brain plays in this interpretation is that we can be aware that we are conscious, which is not necessary for the interpretation to be valid. I cannot prove to you that I am aware that I am conscious. For all you know, I may be a Turing Imitation Game machine cleverly programmed to fool you that I am aware I am conscious. Even if I were a Turing machine, I may still be conscious but unaware that I am. This is probably true of baboons, but possibly not of whales.

The conservation laws and symmetries are valid in every history, each will appear to happen in a "physical world". Instead of trying to see this unfold in Many Physical Worlds or Universes, Occam's Razor suggests that the concept of an ultimately real physical world is a delusion like phlogiston (a mythical chemical element of yore).

Three things Occam needs to cut off the present commonly accepted interpretation of the quantum mechanics and discard, are:

1. Process I of von Neumann. What it explains is already implicit in Process II
2. The collapse of eigenstates upon the entry into awareness of one of them
3. The random selection of the eigenstate experienced, called the Uncertainty Principle, implying the demise of causality

These three beliefs are pure metaphysics. They reveal nothing. The only purpose they serve is to entrench us in delusion.

Number 3 above is metaphysically tangled with the non-commutativity of the operators of observation of complementary variables which must **not** be thrown out with the metaphysical bath water.

Suppose a system is in a state $a.e_1 + b.e_2 + c.e_3$. Where e_1, e_2, e_3 are the complete eigenstates of some observation operator. If we observe the system repeatedly, the number of times we find it in each of these states will be in proportion to the numbers aa^*; bb^*; cc^*; where the asterisk * means the complex conjugate.

Of profound significance in the interpretation of the quantum mechanics is understanding, in the light of the above, why the

following is true: If the apparatus is set up so that a single event at some space-time point can be caused by any of the eigenstates, the amplitude for the event to happen is the sum of **all** the computed amplitudes of all the eigenstates that could have caused the event. Namely a+b+c in this case. There is now no discrimination in favor of one eigenstate and against the rest. This condition is called "interference" because the amplitudes, which are signed numbers, can add to zero. In this case the event, at the point in question, will never enter awareness, even though any one of the eigenstates with non-zero amplitude could have caused it: In other words, all of the eigenstates contribute to what is observed, or not observed. This is the point of departure of a multiplicity of metaphysical interpretations of the quantum mechanics. These are described in a later chapter.

It may seem that a prop to support Everett's formulation is the assumption of the simultaneous reality of all eigenstates. Proof of this reality beyond any doubt is quantum interference. If all states can cause a single event, its amplitude is the sum of all wave-functions caused by each eigenstate.

Instead, if one's awareness is caused by only one eigenstate, the prop, needed to sustain the Copenhagen Interpretation, is the need to keep all the others until the instant one is aware of one of them, then at that instant collapse all the others. *Do this instanter! Don't listen to Albert. He starts kvetching every time we have to do this, but he doesn't understand that we can't let the Special Theory of Relativity get in the way of the preservation of our homeland, the Physical World.* What we are collapsing is the delusion that if they are real, we would be aware of all of them them. Later, we will identify reality with consciousness which is is not observable and awareness with what is observable. These terms will be defined.

This is a paradox trap when we see an indivisible particle, even a whole atom of silver, "smelling out", as Feynman put it, all of the paths to get to its destination, even paths separated by meters, or even light years, and arriving there with the sum of its amplitudes to take each path. And not being able to arrive at all if this number is zero. This defies intuitive reasoning.

If you can only get to work taking either A or B street, you cannot get to work if both A and B are blocked. Obvious. Now take this dose of quantum mechanics: You cannot get to work either if both A and B street are open, but can using A if B is blocked, or can using B if A is blocked. If you took one street, how in heck can the street you didn't take matter? How does reality even know it was blocked if you didn't take it?

Feynman explains: It got "smelled out" and if was open you cannot get to the plant at the corner of Industrial Rd. and Corporate Ave. but others can easily get to work further down Industrial where it crosses Business Way. If this leaves you scratching your head in bewilderment, you are in good company: It left Einstein, Heisenberg, Dirac, Born, Pauli, Bragg, Compton, Bohr, Planck, Skłodowska-Curie scratching theirs. It was not easy for this Nobel Prize winning lady and these gentlemen to give up believing in the ultimate reality of the physical world, and none of them really did. What hope is there for us mortals to understand such matters? It is actually very easy to understand: If the particle can do 911 different things, it <u>always</u> does <u>all</u> of them. If believing the "physical world" is ultimate reality, makes it hard to understand how, it is time to put the "physical world" out to pasture.

Feynman on amplitudes, circa 1960

In his book on undergraduate quantum mechanics Lectures on Physics in the 1960s, Prof. Richard Feynman gave his undergraduate students the following laws to compound quantum mechanical amplitudes knowing the which paths are open to a particle, like a photon in an interferometer. These laws encapsulated what physicists had learned when Feynman was ten years old.

Law 1. The probability of an event is given by the square of the absolute value of a complex number c called the probability amplitude: In practice this is done by multiplying c by its complex conjugate c* which is just c in which the sign of i, the square root of -1, is changed.

Law 2. When an event can occur in several alternative ways, but you cannot **know** which, the probability amplitude for the event is the sum of the probability amplitudes for the event to happen each way considered separately. This is called interference, because amplitudes are signed numbers and can cancel or reinforce.

Law 3. If an experiment is performed so that one in principle can **know** whether one or another alternative is actually taken, the probability of the event is the sum of the probabilities for each alternative. There is no interference, because the probabilities are always positive real numbers and never subtract.

It appears that nature obeys one of two laws, but decides which one to obey depending on what you can **know**. There is no parallel to this in classical science where you can stand outside the system you observe. If you have an apparatus that exhibits Law 2, you can only see Law 3 if you change the apparatus, so it is a different experiment. Of course this is also true in classical physics but we can design experiments so nature seems to do her thing without regard to what you know or don't know. We are surprised when switching from Law 2 experiment to Law 3 experiment that anything we were looking at changed, and even more surprised when physics makes us use different math to represent our experiences. The switch made an enormous change to Alice, whom you will soon meet with her friend Bob, when they were doing Mandel's experiment which we talk about later.

Von Neumann understood this and developed an approximate theory of the interaction of the observed system with the observer, as a quantum mechanical system, to account for the subjective experience of the observer. He was still straddling the classical fence between the observed and the observer. A quarter century later Everett showed that von Neumann's approximation, like all approximations, misses the truth. Everett treated the observed system and the observer as a single system and showed that the mathematics of the quantum mechanics as developed by its pioneers reveals something very interesting about this combined system. We will get to that later after clearing the terrain of metaphysical pitfalls.

Classical physics assumes that when you observe something, although you have to interact with it to do so, you can make the effect of the interaction arbitrarily small, thereby enabling the observation to be made with arbitrarily great accuracy. The quantum mechanics showed that this assumption, generally, is fundamentally flawed. Observation changes the system state by an irreducible amount related to Planck's constant. It also changes you in ways more fundamental than your acquisition of information in the classical realm. In classical physics you can make yourself aware of the state of the observed system with minimal effect on the system: you are a detached impartial observer external to the system. You are changed by the observation only to the extent of becoming aware of its state in a way that minimally affects the observed system. In the quantum mechanics, you and the observed system are one: Until Everett in 1957, you could believe that you were still an impartial observer but your observing the system made unpredictable changes to it. We will show these interpretations to be metaphysics originating in unquestioning assumption of what its real often on the part of those who treat discussion of "reality" in science with disdain. We defer discussion of reality to follow the chapters on the metaphysical views of the quantum pioneers and the chapter on consciousness.

Below we discuss two metaphysically motivated ploys due to Einstein and Feynman to find out "which path" a particle takes in an interferometer. The very question *which path?* assumes that the metaphysics underlying the question is meaningful: by assuming that if it took one path it could not have taken the other because it is an indivisible particle, not a wave, some of which can go one way and the rest go another.

Albert Einstein's ploy: Suspend the slit frame of a Young's interferometer without friction. A photon going through the upper slit will give the slit frame some upward momentum, and downward momentum if the photon went through the lower slit. The change in motion of the slit will tell "which path". Einstein did major work investigating Brownian motion in which a joggling of fine particles seen through a powerful microscope due to random impacts from the thermal motion of nearby atoms is seen. So his idea here was not total fantasy.

Richard Feynman's ploy for electron interferometers: Put a lamp behind the slits that blocks neither slit but fills the space behind each slit with light, so that electrons coming through the slits scatter photons, revealing "which path" by Compton scattering of light behind the slit a diffracted electron presumably passed through.

Feynman analyzed both his and Einstein's ploys using pure quantum mechanical math showing that the more effectively you can find out which path the particle took, the more you destroy interference. In the limit of knowing exactly which path it took, there is no interference.

Metaphysics: Interpretations of the Quantum Mechanics

A source of metaphysical confusion is that the theory predicts amplitudes for several events to happen but apparently does not uniquely predict which one will be observed next, except that events with zero amplitudes will not be observed, so that the prediction is certain only if it is the only one possible. The theory is only known to be correct if a large amount of observed data matches the observed Born probabilities to be in observed eigenstates. This has lead the most astute physicists to conclude that the quantum mechanics, unlike past exact sciences, no longer makes determinate predictions as did the exact classical sciences. This was the conclusion of Heisenberg, Dirac and later von Neumann, then mostly everyone else including Feynman. These views persist to this day. Although Dirac disavowed interpretations of the quantum mechanics, all of its founding fathers, with the possible exception of Einstein, Schrödinger and Bohr, were trapped in a metaphysical box for the next quarter century. Even after Everett then opened it in 1957, many refused to leave. Einstein got trapped in a different box of his own creation for the rest of his life.

Copenhagen Interpretation, 1927

To replace what is now called classical physics, which had been seen as the only rational description of mechanical reality, a new interpretation of mechanics emerged. Heisenberg called it the Copenhagen Interpretation of the quantum mechanics, no doubt to honor Niels Bohr for whom Heisenberg worked at the university of that city. More than a decade later, the Copenhagen Interpretation was presented by Heisenberg at the Solvay Institute conference in Brussels in 1927. It was attended by thirty-odd major contributors to the quantum mechanics. See their photograph in Appendix H. Heisenberg got the Nobel Prize in 1932 and Erwin Schrödinger and Paul Dirac won it in 1933 for their contributions to the quantum mechanics. However, Schrödinger and also Albert Einstein, who got the Nobel prize for explaining the photon, did not embrace the Copenhagen Interpretation.

They were not alone. Even Bohr of Copenhagen had different views than Heisenberg.

The Copenhagen view is that the system state, called the state vector of Heisenberg's interpretation, evolves deterministically in time as a linear superposition of eigenstates: The eigenstates do not change: The computable amplitudes, which multiply them in the linear superposition, change with time. **However, when the system is observed, one is aware of only one of its eigenstates.** This interpretation holds that its *Principle of Uncertainty* prohibits you from knowing in advance what the observed state will be unless it is the only state. But the evolution is deterministic until the exact moment you observe again. Then, suddenly, it appears again in just one state again, possibly different from the last.

Q: Who decides which state?

A: God.

Q: How does He do that?.

A: Plays dice !

At least that's the way Einstein saw Heisenberg explain it. Einstein objected: "I don't believe God plays dice with the universe".

Q: What happened to the others states when the observation was made?

A: Their eigenstates **collapsed,** instantly and everywhere at the instant you make the observation. Superluminal causality, you see: Spooky forces radiate out at infinite speed from the point of observation hunting down unwanted eigenstates and kaiboshing them. Einstein didn't like that either: "I don't believe in spooky action at a distance" which he saw as a violation of the Special Theory of Relativity which would lead to inconsistency in causality.

If then you observed the system to be in state X, it will continue to evolve deterministically starting from X and X alone, but other states may start appearing in time in a deterministic evolution. Until, that is, you make another observation, when all will collapse again except this time to, say, Y. The system then will be only in eigenstate Y. And so on. However, the observer lives in a place called the **classical realm** where the classical laws of physics obtain. There is no uncertainty in making measurements, like pointer and scale readings. This was the Copenhagen Interpretation.

"EPR" Interpretation, 1935

Albert Einstein and his Princeton colleagues Boris Podolsky and Nathan Rosen, known collectively as EPR, published a paper, which some see as philosophical, which sought to show that the Quantum mechanics is incomplete. Although this paper described no experimental apparatus, it is often presented as describing an experiment based on known observations of two particles remote from one another, described as being *entangled* because they originated at a common event. I have not found online a paper by Einstein describing such an experiment but I am not a scholarly historian and will assume it exists somewhere. I will call it the EPR experiment.

As it was seen at the time of the 1927 Solvay conference, the Copenhagen Interpretation would require instantaneous causality: Observing one particle would instantly affect observations made of the other, however remote it may be. EPR accepted that the Copenhagen Interpretation made correct predictions, but at the price of violating the concept of "physical reality": Causal locality was violated, permitting causality to be transmitted at infinite velocity, in violation of Einstein's Special Theory of Relativity, which limited causality to the speed of light. These concepts could be restored, EPR argued, only by recognizing that the Copenhagen Interpretation was incomplete. They used the term "physical reality" as though it were a forgone conclusion to all that it was, like God to the Pope, beyond question.

Bohr did not agree with EPR. Don Howard's *Revisiting the Einstein-Bohr Dialogue* reveals the tenacity with which Einstein, into old age, brushed off any interpretation of the quantum mechanics that did not conform to his preconception of "physical reality" as a way of understanding separability of entangled states. Later we will give reason to believe that there is no such place as "physical reality". We cannot give "physical reality" the brush off and ignore it as Howard points out: *But then the burden falls upon the defenders of quantum mechanics to explain how, while denying separability, quantum mechanics nevertheless rests on an objective ontological foundation.*

We will show that the objective foundation is bare bones quantum mathematics seen as the structure of awareness by Everett, and the ontological part of Howard is consciousness which is incontrovertibly real and needed only in the interpretation, not for what it reveals, but because it blocks any other pseudo reality like "the physical world". Detailed discussion of consciousness and awareness is given in the next two chapters.

EPR argued that something else was going on behind the scenes of the quantum mechanics and it was very simple: Particle pairs, born together like identical twins, had identical attributes, "elements of reality", that predetermined how they would appear when observed. A different pair of twins may have different but identical attributes, EPR said, and both would behave in exactly the same way, but appear differently than other pairs. We don't now know how the particles got these attributes, and the quantum mechanics does not tell us how, or even what they are, EPR admitted, but there is no reason to believe we will not be able to find out.

The Copenhagen Interpretation provided no theory accounting for these attributes, so it was not a complete theory, said EPR. It jumped the gun by claiming to be the last word and paid the price for doing so by needing dice-playing-gods and spooky metaphysics to bolster it.

The notion had been considered earlier that there may be "hidden parameters", referred to above as attributes, not revealed by the quantum mechanics, but which influenced the outcome of observations. These were discussed by von Neumann in 1932.

Einstein never did find out what he said was possible: There can be no doubt that he tried.

John Bell throws a wrench into the EPR works

EPR evidently believed that doing the experiment he proposed would **not** give results that differed from the Copenhagen Interpretation predictions. EPR saw theirs as a philosophical view simply to illuminate an unseen foundation of the theory which would remove its incompleteness. Einstein had amazingly pellucid insight into alternative physical theory. For this reason it is all the more surprising that he did not see that the EPR hypothesis made experimental predictions different from the Copenhagen Interpretation, as John Bell of CERN showed after Einstein died. This implies a difference that is not just one of philosophy, but of science: One could decide between the two hypotheses experimentally. Had Einstein seen this during the remaining two decades of his life, I do not doubt that he would have performed some form of the EPR experiment to resolve the issue that it raised. He was not an experimental physicist, but he was in a very prestigious position, and could have secured the funding and cooperation to do it. It would not have been an easy experiment with the state of electronic art at the time.

If the EPR experiment were done with photons, it is known that two photons can be generated together that are described by the same quantum mechanics wave function, but each traveling in opposite directions. If A, always called Alice, wearing polarized sun glasses sees one photon then B, always called Bob, will see also its mate always as long as the polarization axes of their polarizers are exactly parallel (for simplicity of description we are glossing over a 90 degree phase difference). We imagine that the photon generator is halfway between Alice and Bob who are very far from one another. If Alice's polarizer stops a photon, Bob's will stop its mate. If the photons are generated with circular polarization, there is a 50-50 chance that the photons will be seen or blocked, but the correlation of what Alice and Bob see always will be exact. Experimentally, it is known that if Bob rotates his polarizer by 45 degrees to Alice's about the line of sight, there is no correlation at all between the photons Alice sees with those that Bob sees: If Alice sees one, as often as not Bob's polarizer will block its mate. If Bob rotates his polarizer 90 degrees, there is exact anti-correlation: If Alice's polarizer passes a photon, every time Bob's will block its mate, and conversely. We put Alice and Bob very far apart so that when they see the same thing which hypothetically is caused by some link between them, it would transfer information much faster than the speed of light which is prohibited by the Special Theory.

Later we will show that Alice and Bob agree because, if their polarizers are parallel, the photon is resolved onto the same eigenvectors of the polarizers of both Bob and Alice. They are in the same states of awareness with respect to the event that emitted the photon pair. Both states of awareness due to the two states of polarization are equally real in consciousness, but these states of consciousness does not bind in awareness. It would be premature to do that now – we are looking at the way EPR saw things.

What John Bell did

Bell's reasoning, which Einstein overlooked, is not obscure: Suppose the generator created 100 pairs of photons while the axes of Alice and Bob's polarizers were exactly parallel. But, while the photons were still in flight, Bob quickly rotated his polarizer through a small angle before any photons arrived. If Bob and Alice each had observing apparatus that printed 1 if a photon passed and 0 if it was blocked, then both would have a list of 100 1s and 0s, like 10110001010110…., after all the photons had arrived at Alice and Bob's stations. The disagreement between their lists would be greater if the angle Bob rotated his polarizer was greater, up to 45 degrees when an Alice 1 would match a Bob 1 as often as a Bob 0 and likewise an Alice 0 would match a Bob 0 as often as a Bob 1.

Suppose the small angle Bob rotated his polarizer made a 10% difference between Alice and Bob's lists, i.e. 90% of the photon pairs agreed. For example:

`Alice` 10110001010110100111100**1**00010101100101**0**1…
`Bob`   10110**1**010101101**1**0111100**0**0001010111010001…

With disagreements shown in **bold** type.

Here is the crux of Bell's argument: If EPR are right, then if Bob had not rotated his polarizer, his list would have agreed exactly with Alice's.

Had Alice quickly rotated hers in the same direction a Bob did and by the same amount, while the photons were in flight, her list would have exactly agreed with Bob's, because both their polarizers would be parallel again. We have printed copies of both lists. EPR proclaim that the photons carry with them identical genetic codes from the moment of birth, so to speak, which predetermine how they will respond to observation, so rotating polarizers could have no effect on them while they are in flight. The photons were born with the predisposition to respond to observation the way they did, according to EPR.

Bell asked this question: What would Alice have seen if, while the photons were in flight, she had rotated her polarizer exactly as much as did Bob, but in the opposite direction, so that the angle between her polarizer and Bob's was twice as great. Her list would disagree 10% with the one she got by not rotating her polarizer, as Bob's disagreed by 10%. Therefore, it is impossible for the disagreement of her rotated list with Bob's to be better than 0 or worse than 20%. But the Copenhagen Interpretation predicts more than 20%. We don't have the printed list that Alice would have had by rotating her polarizer in the opposite direction to Bob's because she didn't do so. However, the EPR premise of predetermination allow us to use what philosophers call Counter-Factual Determinism (CFD): CFD stands us in good stead in daily life: If the insurance company bought us a new house because the old one burned down, we can use CFD to tell us what would have happened if we had not bought fire insurance. We don't have to not buy fire insurance to find out what would not have happened if we didn't. However, we can't use CFD with interpretations of the quantum mechanics that EPR believed were incomplete: We cannot have the results of a quantum mechanics experiment we never did. Bohr made this point in his disagreement with EPR. Then why not? Later it will be evident that there is simple explanation why not, but only after slogging out of a metaphysical mire.

EPR proven wrong

The EPR experiment eventually was done by Prof. Alain Aspect in Paris in the 1980s and proved EPR to be wrong. Aspect's results were consistent with the Copenhagen and several other interpretations including that of Everett. Although the Einstein-Podolsky-Rosen (EPR) interpretation was proven wrong, it does not mean Einstein was wrong about God not playing dice, because the EPR interpretation is not the only way of getting rid of the metaphysical part of the Principle of Uncertainty. We are referring here to the metaphysical view that the quantum mechanical future has an element of irreducible unpredictability, not to the non-commutativity of operators. This distinction is discussed later.

The results of Prof. Aspect's experiment and hence the quantum mechanical predictions cannot be explained by classical physics, but could have been had Aspect's experiment proved EPR right. The *Physical World* is manifest to us in two ways: 1. Intellectually in our knowledge of the laws of physics. 2. Intuitively in feeling that we are in such a world. We can give up believing that the classical laws of physics are right. However, doing so does not make the intuitive awareness of being in a *Physical World* go away: We did not and cannot give up our central nervous system: It goes on making the $h=0$; $c=\infty$ approximation it was designed to.

Bohm - de Broglie Interpretation:

A "pilot wave" described by the system wave function evolves deterministically and guides particles to the point of observation which, however, is unpredictable, unless one is aware of the wave-function of the entire universe. This interpretation recognized the potential reality of unobserved states. It recognized the state-vector of the whole universe. In these regards it presaged Everett. It pays lip service to the concept of the physical world.

The Relative States Interpretation, 1957

Hugh Everett III wrote a revolutionary paper showing that the mathematics of the Quantum Mechanics supplies its own interpretation. He was a student at Princeton at the time. His thesis adviser, the eminent physicist John Wheeler, suggested that Everett condense the paper and present it as a PhD thesis; Everett did and was awarded that degree.

Everett saw something in the established mathematics of the quantum mechanics that had not been seen even by von Neumann. It revealed a very different reality than everyone else had supposed. It was not an addition of new metaphysics, not a new interpretation, but something implied by the math from its earliest history. In this respect a discussion of it does not belong in a chapter beginning with the word *metaphysics*. However, every interpreter of the quantum mechanics believes that his view of reality is the "real" physics and does not believe it to be metaphysics. To be fair, I have not put Everett on a *Mine is the Only Reality* pedestal in this chapter. I will do so in a later chapter. Even as I write this sentence, major universities across the world present the

Copenhagen Interpretation as "reality".

Giving enthusiastic support to Everett at first, Wheeler later withdrew his endorsement saying that Everett dragged too much philosophic baggage into the quantum mechanics. Everett's work was popularized by Profs. Bryce DeWitt and Neill Graham in their valuable book *The Many-Worlds Interpretation of Quantum Mechanics* which reprinted both Everett's original unpublished paper and his thesis as well as papers by DeWitt, Graham, Wheeler and others who concluded as had Everett.

My purpose is to give the reader a basis for understanding the predictions of the quantum mechanics and to show that most quantum mechanics interpretations are loaded with metaphysical baggage that make its predictions bizarre and paradoxical. If philosophic baggage was dragged into the quantum mechanics, it was being done by Wheeler and not by Everett. Even the words *Many-Worlds* and *Many-Universes* drag in such baggage unless we look carefully at its semantics. *World* is a word Everett did not use. *Reality* is a word he did not use, except in a note added in proof to his thesis in response to those who had read the proof and did use it in criticism. He did refer to the subjective experience of the observer which implies consciousness at least for human observers.

Everett has been criticized for not elaborating on his interpretation. However, further epistemological elaboration is not needed. He seems to have shunned supplying it with an ontological foundation, perhaps because he felt doing so was unnecessary, or saw no clear path to doing so. The subsequent sixty year history showed that his formulation cannot stand without such a foundation.

When reality is seen as consciousness rather than the physical world of classical physics, Everett's interpretation gives a very simple interpretation of quantum mechanical mathematics stripped of the metaphysical burden of Copenhagen with it random and superluminal causality. This interpretation had to await experiments by Prof. Leonard Mandel in the 1990s, described below, which blew down intuitive metaphysical barriers to understanding. Nevertheless, in a paper by DeWitt reprinted in the above book, he concisely characterizes the key points of Everett's interpretation:

1. No metaphysics need be added to quantum math
2. No observers in a classical realm are needed
3. The entire universe can be characterized by a state vector

4. State vectors never collapse so the reality as a whole is deterministic.

5. The statistical properties of observations are inessential to the foundation of quantum math: They need not be imposed a priori, but can be explained.

Number 3 above was an idea of von Neumann.

The Basis of Everett's formulation

John Von Neumann in *Mathematical Foundations of Quantum Mechanics*, p351 of Beyer's English translation, describes the evolution of a quantum mechanical system by one of two processes:

I A discontinuous change brought about by observation of a state Ψ which changes it to one of its eigenstates, say ϕ_i probabilistically with a computable probability that is a function of the time. If the same experiment is repeated, at the same elapsed time it may appear in eigenstate ϕ_j, who can say and who can tell? So the process is inherently random unless ϕ_i is the only permitted state.

II A continuous deterministic change of an isolated system in which its state Ψ changes at the rate $\partial\Psi/\partial t = A\Psi$, where A is a linear operator.

From the earliest days of the quantum mechanics, this schizophrenia gave rise to metaphysical musings starting with Schrödinger's cat in a box, with the cat in a linear superposition of the states of being alive and dead at the same time until the box was opened.

Everett showed that the quantum mechanics can be completely understood in terms of process II alone as a complete theory. It makes sense to talk about the state vector (wave function) of the whole universe, which evolves deterministically forever. There is no random causality (as a metaphysical interpretation of non-commuting operators). There is no collapse of the system state vector. The observer is seen as a quantum mechanical system that interacts with the quantum mechanical system that is being observed. The observer is not seen as an element of a fictitious classical realm. This interaction is not unique: Each eigenstate of the observed system leaves the observer, upon interaction or observation, in definite states of awareness: Memories, photographs, pointer readings, etc. However, the orthogonality of the eigenstates prohibit any observed state caused by them from containing any memory, etc, of what happened in the others. All of the observer states are equally real. We will show later that this orthogonality makes it impossible to remember the future even though the quantum mechanics itself is time symmetric.

It is easy to see that the simultaneous reality of contradictory states of awareness cannot be explained as a "mind" phenomenon because awareness may never enter the human brain. In any case it can have consequence of great magnitude in what we see as the external physical world. Everett's formulation can only hang onto the physical world by going to a Many Physical Worlds interpretation which has absurd consequence. One is left trying to understand what the words *mind* and *world* can possibly mean in the light of Everett's formulation, which is described in the chapter *Quantum Mechanics Interprets Itself*. The purpose of this book is to show that the impasse in finding the ontological foundation of the quantum mechanics is the belief that it must be grounded in the physical world.

The reality of observer states is not measured by their amplitude: Reality is tagged by vectors in Hilbert space. An analogy: Winning first prize in a lottery. You cannot win first prize a little bit. You either win completely or you do not win it at all. Winning is real, or it isn't. However we can talk of the density of winning states. If I bought 3 tickets and you bought 30, the density of your winning states is 10 times greater than mine: In the long haul, you will win ten times as often as I. My experience of winning, although it seldom happens is just as real as your experience of winning which happens to you more often. This density is related to the amplitudes in the quantum mechanics: They are not measurements of reality. There is no physical world analogy, hence classical physics analogy, that illustrates the quantum mechanical properties of reality. However, it is easily understood in terms of consciousness which we all experience. We take up this point at the end of the chapter on consciousness and awareness.

Everett's formulation creates serious problems if reality is equated with the physical world of classical physics: Every quantum event causes the physical universe to spawn multiple, generally infinite, numbers of universes with consequent multiplication of energy content. This evidently was too much philosophical baggage to accept for Prof. Wheeler, Everett's thesis adviser. When one realizes the notion *the ultimately real physical world* is itself the baggage we need to leave behind, this paradox vanishes. At date, six decades have passed since Everett's thesis and as far as the major universities go, it is all but forgotten. Why? One <u>has</u> to talk about **reality** to understand why:

People cannot stop believing in the ultimate reality of the physical world. DeWitt and Graham's book popularizing Everett's work is titled *"The Many Worlds ....."*. Not only did it <u>not</u> bury the physical world, it exhumed infinitely many of them, and attributed them to Everett.

<u>Subsequently in this tome, two words will be defined:</u> *<u>Awareness</u>* <u>and</u> *<u>Consciousness</u>* <u>Within the context of these definitions we will identify Awareness with von Neumann's Process I and Consciousness with his Process II.</u> We will show that awareness is the binding of consciousness. This binding has its origin at the most fundamental level of the quantum mechanics. The mathematics of awareness was given by Everett.

Giving Everett's mathematical analysis here or in the chapter on quantum mathematics is an invitation to fall into traps of interpretation that have snared the greatest thinkers of quantum mechanics. It is deferred to the chapter entitled *The mathematics of quantum mechanics supplies its own interpretation,* after the chapter on consciousness and awareness.

The Many Minds Interpretation

This interpretation jumps out of the frying pan of the Many Worlds interpretation of infinitely many physical universes incurred by each quantum event, a metaphysical view troweled onto Everett's formulation by DeWitt and Graham. I do not know what the word *mind* means without a definition of it in terms of what later I will call consciousness and awareness, so I cannot understand it.

Feynman and metaphysics

Richard Feynman in this three volume tome for undergraduate students, **Lectures on Physics** expressed a view that few physicists in 1960 would dispute: That physics has given up on the problem of trying to predict exactly what will happen because it is impossible to make predictions that are anything but randomly determined – the antithesis of determinism. He saw this as a retrenchment in the ideal of understanding nature. A backward step, but he believed no one has seen a way to avoid it.

In 1964, Feynman in Lecture 6 of the Messenger Series of lectures at Cornell, available on-line, described a particle interfering with itself by taking both paths through a two slit interferometer to a point on the screen where the amplitudes to arrive there add as signed numbers. They can add constructively or destructively giving interference fringes. Yet if you try to find out which path the particle takes you will destroy the interference fringes. If you look behind both slits, the particle only takes one path or the other – never both. Its choice is fundamentally random, he says. Feynman made the profound observation that this randomness is so fundamental that it cannot be described as ***stochastic***, which implies that some path is known to God but we do not have the theoretical ability to know which path God has chosen, nor ***statistical*** which implies that we do know the theory but do not know the initial conditions well enough, or at all: Because, if the path is known, even by God, there will be no interference. Feynman put it this way: So even God does not know. In this lecture Feynman gave this famous advice: *Do not keep saying to yourself, if you can possibly avoid it, 'but how can it be like that?' because you will get 'down the drain,' into a blind alley from which nobody has yet escaped. Nobody knows how it can be like that.* Notwithstanding this profundity, we will show that understanding how a two slit interferometer works is so easy even a schoolchild can do so, provided they leave their Physical World toy outside the classroom.

In 1979 Feynman gave a series of lectures at the University of Auckland in New Zealand (available online) in which he firmly described photons as particles because that is how they are observed. He seemed to shun the use the word wave in describing their propagation. Their behavior was unpredictable: A statistical averaging was needed. When asked whether he thought the future would bring change to this picture, he replied he could see no possibility that it would, but made no reference to Everett, whose work two decades earlier Feynman could have seen without a crystal ball.

Remarks on Feynman's view

It is hard to believe that Feynman, who expressed these views in his 1960 book, had not read Everett's thesis written a few years earlier. They were both graduates of Princeton and both had the same thesis adviser: John Wheeler. It is not idle to speculate that Feynman did not think about the possibility that the above views were metaphysics, let alone see that they were the very metaphysics that Everett's thesis undermined.

The only way I can understand Feynman is to assume that he took the same view as von Neumann in believing in the ultimate reality of the physical world which would require most quantum events to spawn multiple or infinite numbers of physical worlds in Everett's formulation, which thus could be rejected out of hand as absurd.

The purpose of this epistle is to show that exactly the opposite is true.

The quantum mechanics has made it possible to predict exactly what <u>will</u> happen, not is a metaphysical "physical world", but in consciousness. Because it does not exist in a physical sense, infinitely many states of consciousness is no more of a burden to God than infinitely many numbers. Feynman, and everyone else in his time, took the quantum mechanics as serving the same purpose as classical science: To predict awareness: We are aware of things happening in the physical world which is taken for granted to be ultimate reality. We cannot believe in multiple inconsistent physical worlds: We would be aware of all of them if they were real.

Exactly what I mean by *consciousness* and *awareness* will be given in more detail later. It will be shown that blurring the distinction between them gives rise to absurd metaphysics when extended beyond classical physics which assumes one will be aware of whatever is real. In classical physics, there was no need to make a distinction between consciousness and awareness, although there was no reason not to. In common parlance, these two words often are used interchangeably.

Leonard Mandel, 1992

In the 1990s experiments were done by Prof. Leonard Mandel at the University of Rochester to investigate claims that quantum interference caused by an event which can happen in multiple ways by splitting the light (or other particle) along different paths, is destroyed when the observer interacts with the beam paths by trying to determine which path the particle is taking. Using theoretical reasoning, Feynman showed that probing the beams to determine which path the photon takes in a two path interferometer destroys interference to the extent that the probing is successful in revealing "which path". I put the words *which path* in quotes because those very words embody the metaphysical assumption that the particle, being an indivisible particle in a physical world, can only take one path or the other but cannot take both. Mandel designed an interferometer in which he could determine which path the photon took without any probing of the beam paths of an interferometer showing interference fringes. His results were described by some physicists as "mind boggling", even though the results were completely predictable from the early pages of Prof. Richard Feynman's book on the quantum mechanics for undergraduate students, written thirty years earlier: *Lectures on Physics, Vol III*. The rule was given that interference would be destroyed if one can somehow be aware which path the particle took, which Mandel proved without any probing. Minds were being boggled not by physics, but the paradox of disagreement between their metaphysical belief and reality, while presuming that their metaphysical belief **was** reality.

Mandel's work is covered in a later chapter. Because of the need to clarify what is known about consciousness in order to interpret Mandel's results, that chapter follows the chapter on consciousness.

The Fall of Classical Physics

This quagmire of metaphysics is best understood in the light of the history of rise of classical physics, including Euclidean geometry and kinematics, which stood on a foundation of dry, solid adobe. The structure was sound and understood as consistent with intuitive models forged in our brain by the hammer of experience over eons, and wired into our central nervous system. This intuitive model makes three implicit approximations: The velocity of light = ∞ (infinity); Planck's constant $h = 0$; the curvature of the earth's surface = 0. These approximations are very accurate in the domain of ordinary experience. Intellectually knowing the correct values of these constants does not deliver us from the intuitive conviction that we live in what I will call the $c = \infty$; $h = 0$ world, although most people now are delivered from the belief that the earth is flat. The $c = \infty$ approximation separates space and time and leads to the intuitive view that space, time and mass, the cornerstones of civil and scientific metrology, are absolute attributes of physical objects; that at the present moment causality transmits instantaneously to the boundaries of the universe; in other words, what is now here, is the same now everywhere else. The $h = 0$ approximation has deeper consequence: the intuitive view that a unitary physical universe is ultimate reality.

In retrospect we can see the collapse starting in the early 1800s with the simple experiment that determined an electromagnetic constant of a vacuum to be the velocity of light. The fog lifted in the late 1800s when Maxwell showed that electromagnetic waves must propagate with the velocity of this constant and not at a velocity determined by kinematic considerations: The velocity of electromagnetic waves does not enter into Maxwell's theory via consideration of reference frames in relative motion. By then it was certain that light propagated as an electromagnetic wave.

Then, the Michelson-Morley experiment failed. It was a simple and exact experiment that showed the earth always to be absolutely stationary, which was unbelievable: The earth was moving around the sun at 40km/s and changing direction so regardless of the sun's motion through space, the change in the combined motion of the sun and earth had to be measurable: This reasoning seemed beyond question to everyone, so the results of this experiment was a scientific paradox. It is interesting to think of how this delusion originated. Here is a suggestion: You know where you are in your living room, its walls are a frame of reference identifying where you are located in the room.

You know where the room is located in the house;

you know where the house is located on the lot;

you know where the lot is located on the block;

you know where the block is located in the suburb;

you know where the suburb is located in the city;

you know where the city is located in the county;

you know where the county is located in the state;

you know where the state is located in the country;

you know where the country is located in the continent;

you know where the continent is located on the planet;

you know where the planet is located in the solar system;

you know ....

This ad nauseum repetition is intended to convey ad infinitum regression towards a greater and greater framework of reference. They cannot take that away from us, we feel, so it all must end in that Great Frame of Reference in the sky: Space Itself. Intuition. Michelson-Morley were trying to measure the motion of the earth through "space". This was their delusion that the Principle of Relativity killed. What is observable is only determined by relative motions of the observer and the thing observed. It took Einstein to point out this simple truth. "Moving through a vacuum" is a delusion: Moving relative to what?

The marriage of Newtonian mechanics and Maxwellian electromagnetics made absurd predictions: A red hot poker would glow brighter in the ultra-violet than the red. Worse yet the total radiation would be infinite! This was really bad news for the numerous scientists that tried to squirm out of these absurd predictions and could not. They called their predicament the "ultra-violet catastrophe".

The first rains of the approaching storms of the quantum mechanics and the theory of relativity had come and started to turn the adobe foundation to mud. By 1905 the bastion of classical physics started to crumble. Everything sacred, seen as the ultimate correspondence between the mind of man and the mind of God was gone: Geometry;

kinematics; the laws of dynamics; the theories of fundamental forces. By 1930 the classical realm, a once towering sculpture revealing the structure of reality, was no more. Some of its building blocks, with the mud washed off, are to this day excellent tools for engineers and architects and many sciences. The mud has not yet been washed off beliefs that were held dear before the rains. The purpose of this paper is to show that these beliefs were more than covered with mud, they were the mud which, as adobe, seemed to be the rock solid substance of physical reality. The metaphysics of classical physics is deeply ingrained in common sense and intuition that the thing external to consciousness, the physical world, is ultimate reality. It is very hard to shake this ingrained perception. An Apollo astronaut described nausea, overriding his intellect, caused by seeing the earth upside-down. Gut level intuition is deeply rooted.

The founders of the quantum mechanics struggled with the attempt to understand reality in the light of the new mechanics. Von Neumann, who believed the physical world was reality, understood that the quantum mechanics had to recognize consciousness. He did this with the proclamation that consciousness was "in the physical world", and devoted much of chapter V and VI of *Mathematical Foundations of Quantum Mechanics* to model conscious experience.

Intuition and common sense

Nothing in the above should be construed as advice not to trust intuition and common sense in general. They lead us astray in the attempt to understand the results of experiments on individual elementary particles and rarely in understanding macroscopic events. Those who would say: "Oh them electrons. I don't understand that stuff. It's of no importance to me" need to be reminded that they are made out of elementary particles, so there is no escape from them outside the box of ignorance. However, there are two respects where intuition and common sense tell us truth in macroscopic experience that is supported by the quantum mathematics that is not or is very poorly supported by classical physics; truths which are obscured by metaphysical interpretations of the quantum mechanics. These are the arrow of time and freedom of will, discussed in a later chapter.

Consciousness

The Doctrine of Psycho-Physical Parallelism

The philosopher and mathematician Gottfried Wilhelm von Leibniz, a contemporary of Isaac Newton and co-inventor with Newton of the Infinitesimal Calculus, presented us with the Doctrine of Psycho-Physical Parallelism: Reality always presents itself to us in two very different ways:

1. in consciousness
2. as the physical world external to consciousness.

If you are a scientist who thinks *doctrine* is a dirty word, substitute *principle*.

Example: You are sitting at a table once owned by your grandfather. You consciously feel the visual sensation of seeing the smooth table surface and its edge and a corner. You place your hand above the surface and consciously experience the visual sensation of seeing it there. You lower your hand and, when you see it touch the surface, a new conscious sensation arises: You feel a smooth hard surface touching the palm of your hand. You run your hand to the edge you can see, and you can feel the edge. Running your hand along the edge you feel the corner where you see it to be. If you rap your knuckles on the table surface, you feel a sharp sensation in your knuckles when you see them strike the table op and at the same time a new conscious sensation arises: the sound of knocking on wood. You are not inventing all this. Others describe having the same conscious experiences with the same table.

You get up and leave the room. You don't believe the table ceases to exist when you no longer consciously experience it. Your conscious states you experienced arise from the activity of atoms in your brain which links through senses to the table. The table and your brain exist, in some sense, independently of your conscious experiences of it and the table and atoms of our brain existed even before you were born.

Since Leibnitz' time, most philosophers agree with this doctrine, but divide into two schools:

    A. A bridge between consciousness and the physical world does not exist: The schism is fundamental.

    B. There is a bridge: The brain.

School A is not deterred by School B: A neurosurgeon operating on your brain and measuring you neural firing patterns will never encounter your conscious states. The operation can be done while you are awake; the surgeon may find that you report feeling various sensations when your brain is stimulated in different ways, but has no way of knowing what it feels like to be you, except to be you, and no scientific measurement can describe it. A surgeon operating on their own brain may know what it feels like and may know how the stimulation produces scientifically describable patterns of neural firings, but cannot convey these feelings in print in a scientific journal, but they can describe the associated pattern of neural activity scientifically. The best that can be done is to appeal to the experience of the recipient of a similar experience. Electrical stimulation of the brain can result in bizarre conscious states that are not scientifically communicable. What does it feel like to be a bat flying in the dark and "seeing" its prey with sonar? Even if you have a scientific description of the neural firing patterns in the bat's brain you still cannot know what it feels like to be a bat. Unless you are a bat. When Dracula bites you, turning you into a bat for a night, when you are human again you won't remember what it feels like because you no longer have a bat brain.

There is a simple reason that science cannot observe consciousness: Consciousness does not **exist** in a physical sense. **But it is absolutely real because we experience it.** We can only observe things that exist. But what is such a thing, like a hammer? We see it as a self-contained ponderable object always occupying a definable place in three dimensional space and always somewhere in that space; it exerts considerable force on nail heads when struck by a fast moving hammer head. We may recognize it as the same hammer Granddad gave to Dad, but with a couple of new handles in its time and one new head. You may see yourself in the same way: self contained and unambiguously distinguishable from everyone else since you were a child even if you have undergone several interchanges of atoms since then. Not to worry: all electrons are indistinguishable from one another as are other particles from their identical twins. The hammer, or you, are focal points of a host of conscious experiences of many people over time but existing independently of any of them. We feel comfortable shifting reality from the pain we felt the time we struck our thumb with the hammer, to the hammer itself, as ultimate reality constructed of hammers, nails, suns, moons and sealing wax. Classical physics is the theory of how that existential reality works. **The quantum mechanics story is different.**

The reality of consciousness

If you don't believe *consciousness* is real, try sitting on a thumbtack. What you feel, not my words, defines what I mean by the word *real*. I am not talking about *awareness*. You may be aware, erroneously, that you sat on a scorpion. Awareness is your understanding of why you are feeling the pain, and it may give rise to conscious states independent of the pain, like the consciously experienced emotion of anger that someone put a thumbtack on the chair. Believing it was a scorpion may give rise to the sense of fear that it may sting you again, being unaware that it was a thumbtack and not a scorpion. Awareness is belief; consciousness is not, although they are often concomitant.

The non-existence of consciousness

To anyone who believes consciousness exists, one may ask to be handed some of it. Where has science ever observed a flask or test tube of consciousness, measured its density, measured its electrical conductivity, or any other measurement of an attribute of physical reality like the atomic weight of lead, the charge on the electron, the pressure of the atmosphere, etc.

If you measure the length of Granddad's table, you can describe it two ways:

- You are comparing an existential tape measure with the existential table top where both exist as physical objects. This measurement can be done by machine. The consciousness it experiences, if any, is unknown.

- You are comparing your conscious experience of a tape measure with the conscious experience of the table top.

Both yield the same number. You can likewise weigh the table or make any other scientific measurement of it. But no one has ever weighed consciousness. It may appear that consciousness can be measured in space and time, but we are only comparing a conscious experience being measured with the conscious experience of a tape measure or the conscious experience of a clock. We can get an observable number without going outside consciousness. If you look in space and time you will never find items of existential consciousness to measure. You cannot reach with existential physical world instruments into consciousness to make a measurement for the reason that consciousness does not exist. But it is real. Skeptical? Try sitting on the thumbtack again.

Science hitherto does not care how you got the numbers. It talks about the mathematical relationships between numbers. Suppose the table was very heavy and you were talking on the phone to a moving company who wanted to know how much it weighed. You have no way of weighing it but you know it is made of mahogany. You can measure its dimensions, calculate its volume and multiply that by the density of mahogany. It does not matter which way you got the measurements: Either method above will do. Science does not care if you think the table exists externally to consciousness or if you believe that only consciousness is real. The scientific theory: volume x density = mass knows nothing about ultimate reality. It is a formalism that represents observed numbers, like pointer readings, and belongs to classical science. **The quantum mechanics does not do this.** Classical physics and quantum mechanics have this in common: The theories are mathematical and exist as such. This is discussed in the chapter: The Foundation of Mathematics.

George Berkeley (Bishop of Cloyne), in the early half of the 1700s, proclaimed in effect that only consciousness is real. It is structured, and such structure may be said to exist, mathematically, but not as a material substance. He described this as idealism. He proclaimed physical reality to be a delusion.

This story is told by Boswell about Dr. Samuel Johnson, of *A Dictionary of the English Language* fame:

After we came out of the church, we stood talking for some time together of Bishop Berkeley's ingenious sophistry to prove the non-existence of matter, and that every thing in the universe is merely ideal. I observed, that though we are satisfied his doctrine is not true, it is impossible to refute it. I never shall forget the alacrity with which Johnson answered, striking his foot with mighty force against a large stone, till he rebounded from it, 'I refute it thus.'

It seems that Dr. Johnson did not hurt his foot. Pity. Had he done so, he may have been more open to Berkeley's point. It is not possible to know what Berkeley meant by *non-existence of matter* without knowing what he meant by *existence*. One may infer that he believed consciousness does "exist". One cannot climb out of a semantic morass by assuming words have God given meanings. For reasons given earlier we have cause to believe consciousness does not exist in the sense we imagine the world of classical physics to exist, but that it is real: Imagine a bad headache. We cannot observe the conscious experience of a headache with instruments of metrology. We can say the headache feels to be in space: inside my head, started at two o'clock and ended at three. However awareness of space and time are also consciously experienced states of awareness. We are not observing consciousness but experiencing it: In principle we can observe the neural firings in our brain that give us awareness of space and time and tie them to the observable neural firing that cause the pain but we cannot pull out the conscious experience of pain out with a hypodermic, squirt it into a test-tube, measure its electrical conductivity, add a few cubic cemtimeters of ectoplasm and observe the precipitate with a microscope. It is a sensation, not a thing like Grandpa's table or hammer: The paradoxes of the quantum mechanics arise when we also bestow on the hammer's existence the belief that it is ultimate reality. We feel our beliefs justified when we can predict correctly the table's weight given its volume calculated from its dimensions and the density of mahogany. We related the pointer in our measuring tape window pointing at length readings to the measurement the moving company got weighing the table and finding it to be 110 kilograms. We can proudly proclaim that we predicted 109.7. **This is not how quantum mechanics works.**

When talking about these fundamental things, it is very easy to end up thrashing around in a mind muddled mess of words, all the time believing oneself to be profound. When Berkeley says the physical world does not exist, he misses a very important point about it: It exists mathematically as a theory that, in a wide domain of experience, is very accurate, but otherwise is completely wrong. The best example of such a theory is that of the flat earth.

Because classical physics did work, we hold that the Physical World is ultimately real which causes ultimately real activity in my brain cells, and consciousness is some sort spin-off of brain function. The contractor who built my house assumed plumb lines are parallel. He used the theory that the earth is flat to build a house that has stood for 100 years. The fact that it still stands is no proof that the world is flat. The fact that much conscious experience is accurately predictable using a model of Physical World does not prove that it is the physical world ultimately real. Both the flat earth and classical mechanics are known approximations. Before they were known to be so, it was reasonable to assume that they were absolutely true theories and the physical world they are talking about therefore is ultimate reality. Immanuel Kant upheld Euclidean geometry as an ultimate correspondence between the mind of man and the mind of God. Belief in the ultimate reality of the physical world extends Kant's idea to all classical laws. This conception of reality should have ended with the quantum mechanics. It did not.

Schrödinger

When the quantum mechanics made its debut, one of its founders, Prof. Erwin Schrödinger, wrote three short essays on consciousness, but did not publish them until his retirement, when he wrote three more, publishing all six in a booklet entitled *My View of the World*. Schrödinger shared the Nobel Prize with Dirac for their wave function theory of the quantum mechanics. I gave my copy of his book away and have not been able to find it again on the used book market. Evidently other or overlapping material with a similar title was published, perhaps later editions, but I have not been able to find exactly the same book. I do see online quotations from it. One of the later essays concerned the ancestral memory of birds: Their memory of how to build a nest comes down to them in their genetic code.

Two of the earlier essays, I believe, are of profound significance to the interpretation of the quantum mechanics. Both originate in the philosophy of ancient India. The first essay described a paradox that Schrödinger attributed to the Samkhya (Sankhya?) philosophers of ancient India. The second concerned a principle Schrödinger attributed to the Vedanta school of Indian philosophy.

Schrödinger was of a philosophic turn of mind and roamed the philosophy stacks of the library of the University of Vienna, his alma mater, where he was exposed to what was reputed at that time to be the worlds best collection of books on Indian philosophy.

The paradox of the Samkhya philosophers

My only knowledge of this paradox is Schrödinger's version of it. I doubt that the ancient philosophers presented it exactly in this way:

You devise an experiment to see if consciousness can be put into a scientific description of experience. You select two subjects who are equivalent for the purposes of the experiment. One could be blind and the other sighted if you experimented with auditory consciousness, but in his example both are sighted. You place one observer at a window through which they can see the view of a garden and the other in a darkened room. You ask both to report to you what they consciously experienced. Then you interchange them. This should make no difference because they are equivalent for the purposes of the experiment. Indeed exchanging them does not make any difference: You get the same results both times. Then, where is the paradox? For no scientifically knowable reason there is one person in the whole world who cannot be a participant in this experiment, or it will fail. In

Schrödinger's words: like one scale pan of an equally loaded balance being weighted down. That person is **you.** The first time you will have a consciousness experience of seeing a garden through a window and the second time you will have a consciousness experience of being in a darkened room, which is completely different. But science tells you: **Nothing changed** when the participants interchanged. **You** are the fly in the scientific ointment. If we can only get rid of **you**, we can put consciousness into science.

At least in the book referred to above, Schrödinger professed not to know how to resolve this paradox. To repeat what I said earlier about paradox: It is always caused by flawed thinking. Paradox is not inherent in reality.

The problem is not with you. What you consciously experienced is exactly what you would have expected. The problem is with science, which acknowledges your different experiences when you go from garden view room to the dark room, if you are only participant. Science says the asymmetry in consciousness will go away if you change places with someone else equivalent. However, introducing the second person does not cancel conscious asymmetry, but doubles it.

The non-scientific reader may better understand the predicament of science by imaging a scientist doing the experiment with two equivalent cameras. They do not have to be identical but should have the same focal length and frame size. You are the scientist. After the first pair of pictures are taken, the cameras are interchanged and the second pair taken. You bring up on your computer screen the first pair one above the other with the garden picture on top and the dark picture below it. To the right of this pair you place the second pair of pictures taken after the cameras were interchanged. All four pictures now enter your consciousness. Your conscious experience of the first pair is practically the same as that of the second pair taken after the cameras were interchanged. You can see no Samkhya paradox. Staring vacantly at the center of the screen wondering why not, suddenly, for some reason convenient to this discussion, the corpus callosum of your brain stops working. The Samkhya paradox returns. The reason is explained in the next two sub-chapters.

While consciousness cannot be observed, the subjects of scientific experiments experience it. If understanding their experience confounds us with paradox, we need to examine the naive beliefs we hold to about consciousness and awareness. An example of such belief about the sequence of subjects: people; baboons; insects; cameras; teacups; electrons; that consciousness stops below people. No, maybe baboons too. Well, OK, maybe insects, but definitely not cameras. Etc, etc. Whatever consciousness is associated with teacup atoms, you may be sure that the teacup itself is not aware of it, aware of itself, or otherwise gives a hoot.

Principle of Vedanta

In the next essay in this little book, Schrödinger described the Principle of Vedanta: There is only one conscious self: You. This principle is stated in the Sanskrit words *Tat tvam asi* translates as *That thou art*. He professed to believe this principle but said there is no reason to believe it. It is a matter of personal preference, he said.

However, there is a very good reason to believe it: **It is needed to resolve the Samkhya paradox.** Why it does is described below, after consideration of experiments that underscore the definition given in this paper of *awareness*.

Michael Gazzaniga

In 1967 there appeared in *Scientific American* an article by the psychologist Michael Gazzaniga on split brain experiments for which his professor Rodger Sperry was awarded a Nobel Prize. Brain injuries are usually fatal, but there were reports in medical history of some freak accidents that were not: A piece of sheet metal, or something similar, fell on the patient's head and without injuring either hemisphere of the brain severed the corpus callosum, a bundle of 150 million nerve fibers interconnecting the two brain hemispheres at the base of the brain. After recovering from this terrible accident, the patient seemed to recover normal functionality and could do very well, thank you, without a corpus callosum. In some cases, such an accident cured the patient's epilepsy. Sperry found that animals without a functioning corpus callosum are disabled to a greater extent the more primitive the animal. A grand mal epileptic who has one seizure every six weeks will likely not allow their corpus callosum to be severed surgically to cure epilepsy, but an epileptic who has several seizures a day has little to loose. Jokes were made about the human corpus callosum that it seemed to serve no purpose but to transmit epileptic seizures from one side of the brain to the other.

Gazzaniga found it difficult to believe that the corpus callosum served no function in humans, and secured the cooperation of someone who had had theirs surgically severed, presumably because of epilepsy, to undergo psychological tests.

Although it may not have been their intention to do so, Sperry's and Gazzaniga's work casts considerable light on the distinction made here between what I have described with the words *consciousness* and *awareness*. The reason for this is that damaging the corpus callosum does major damage to awareness as Gazzaniga showed but leaves consciousness relatively unaffected because it is associated with the cortex, which the freak accident did not damage. I am only using the words consciousness and awareness in the sense I define them. Visual consciousness is associated with the visual cortex at the back of each brain hemisphere, and the optic nerves and eye retinas, which the accident left uninjured. Gazzaniga was aware of the philosophical significance these experiments as I discuss below.

I will only describe two of Gazzaniga's many experiments. The subject stripped to his underwear and was blindfolded, was touched somewhere on his body and asked to point to the place where he was touched. He did so with no hesitation with the hand on the side of his body that was touched, but could not do so with the other hand. If touched on the other side of his body, he could now point without hesitation with the hand on that side: the hand that could not point in the previous test, but could not point with the other hand that was previously able to do so.

In another experiment, with the subject fully clothed but blindfolded, Gazzaniga held the finger tip of one hand on the sagittal plane of his body and asked him to touch it with the index finger of the other hand. The subject could do so, perhaps with less success than someone with an intact corpus callosum. Baffled why he could do it at all, Gazzaniga asked him to remove his shirt. The subject could not oppose his fingers at all without his shirt on.

The high intelligence of each human brain hemisphere evidently allows the hemispheres to establish, from experience, ways of communicating externally to the brain in the absence of a corpus callosum – in the latter experiment the way his shirt felt allows each hemisphere to infer where the arm it does not control is positioned.

Gazzaniga and the Principle of Vedanta

Gazzaniga ended his article with a statement that his experiments prove that the Principle of Vedanta, as Schrödinger called it, is false, or else it would reestablish communication between the brain hemispheres after the corpus callosum was interdicted. To discuss this idea, I need two words and will use these: *consciousness* and *awareness*. I will use *consciousness* as I have defined it above and reserve the word *awareness* for what Prof. Gazzaniga observed.

<u>If the Vedanta Principle refers to awareness, Prof. Gazzaniga has proven it false.</u>

If the Vedanta Principle proclaims that the conscious self is universal, Prof. Gazzaniga's experiments prove that the subject's **awareness** was damaged by interdicting the corpus callosum. Doing has no effect on the consciousness of the subject which is associated with cortical activity of the brain not damaged by the injury. His experiments showed that cutting the corpus callosum had no effect on the *consciousness* of each hemisphere but destroyed the <u>binding</u> of these conscious states in *awareness.*

While Prof. Gazzaniga himself can experience consciousness, he cannot scientifically observe his subject's *consciousness,* because, although real, it does not exist. But he can, and did, observe *awareness.* Gazzaniga can prove scientifically that the subject was aware of where he was touched, but he cannot observe in *consciousness* what it felt like to be touched. His experiments are crucial in making the distinction between the words in italics, although the purpose of them was to determine the effect interdicting the corpus callosum had on what has been defined above as *awareness.* If you object to my semantics, please replace everywhere in this paper these two words with whatever you think I should have used. It is not my purpose to establish semantic convention.

There is a stark distinction between awareness and consciousness in the sense defined above:

- **Awareness is scientifically observable, which is what Prof. Gazzaniga did.**
- **Consciousness is <u>not</u> scientifically observable even if the associated neural activity in the cortex <u>is</u> observed.**

If brain activity is altered by the experiment, the subject is powerless to convey the experienced consciousness to the investigator. If the investigator operates on himself he can know how his actions consciously feel, but cannot convey it scientifically to anyone else.

Prof. Gazzaniga proved that cutting the corpus callosum divides the subject into two disjointed spheres of awareness, (ignoring communication external to the skull learned later). So why not also divide consciousness along the lines of awareness into two conscious selves? There is a compelling reason not to: Doing so results in the paradox of the Samkhya philosophers. Paradoxes are always caused by flaws in our thinking, in this case by presuming conscious self-hood to be denumerable. In Schrödinger's experiment we can place the **awareness** of the subjects into a scientific description: Interchanging the subjects changes nothing. It also changes nothing in consciousness if both subjects are the same conscious self, however dichotomized they may be in awareness. But why not just live with the paradox of the Samkhya philosophers? What have we got to loose? <u>There is a compelling reason for not living with this paradox: It does violence to the interpretation of the quantum mechanics.   It is the cause of the widely held belief that the quantum mechanics is random, debarring causality in theory and the generator of the metaphysical construct of collapsing wave-functions.</u>

It is important to make the distinction between, say, visual awareness and awareness of visual consciousness. If you read in the paper **MARKET CRASHES** or you read 𝔐𝔄𝔕𝔎𝔈𝔗 𝔈𝔕𝔄𝔖𝔥𝔈𝔖 you will be aware in both cases that the market crashed, but your visual consciousness will be different in each case.

Awareness is the binding of consciousness

In Schrödinger's Samkhya experiment, if you recognize that the subject at a window looking at a garden and the subject in the darkened room are the <u>same</u> conscious self, interchanging them changes nothing. There is no corpus callosum interconnecting their brains, so the subject in the darkened room cannot be aware of the experience of the subject at the window, or conversely.   There is therefore no binding of the conscious states experienced by one person and those experienced by the other. **Awareness is the binding of consciousness.** When you, as a participant in the Samkhya experiment, are exchanged with me, there is no change in consciousness to the extent that we are equivalent, which is being assumed. So the reality of consciousness is invariant under the exchange as long as we see consciousness conforming to the Vedanta principle: ***Tat tvam asi.***

Schrödinger and Gazzaniga after brain surgery

In some future age we can imagine an epileptic can have a slice removed from the corpus callosum and replaced with an electronic interface that restores its function.   When the patient has an aura presaging an approaching seizure, they take a remote controller out of their pocket and turn the corpus callosum off. When the danger of the seizure is past they can turn it on again. We do not live in such an age so this is a Gedankenexperiment.

If you have had this operation done to fix your epilepsy, you can repeat Schrödinger's Samkhya experiment by looking through a binocular eyepiece that shows two different images in the right and left visual fields, say a kitten in one and a puppy in the other. These two visual fields are processed by two completely different configurations of atoms of the physical world. The apparent left field of both eyes is imaged on the right side of the retina in both eyes and separate optic nerves go to the right visual cortex at the rear of the right hemisphere of the brain where the images are combined; the right hemisphere also controls the left hand. The converse is true of the apparent right visual field: Just interchange the words *right* and *left* in the above sentence.

Suppose the optics are so constructed that you cannot peek with your right visual field at the left image, or conversely. You are given a thumb-wheel that you can turn, like that on some computer mice. You are asked to take the mouse in one hand and turn the thumb-wheel which will flash various thumbnail images in both visual fields, one of which is the same as the image of the kitten. You are asked to find that image. There are various possible outcomes of this experiment depending on whether your corpus callosum is turned on or off and whether you control the mouse with your right or left hand and in which visual field the image you have to match is being projected. This could be a computer controlled double blind experiment with the computer turning the corpus callosum on and off without the knowledge of the subject or experimenter until the experiment is over. We can predict the outcome from Prof. Gazzaniga's experiments. If your corpus callosum is turned off you will not be able to select the image with either hand unless the thumbnail is in the same visual field as the kitten, but then only with the hand on the same side as the apparent visual field. If it is on, you will be able to select the image with either hand. To the first order, turning your corpus callosum on or off does not change consciousness. The neural activity in your visual cortices is not affected. It is a piece of factitious metaphysics to believe that turning your corpus callosum on or off divides you into two conscious selves: such a dichotomy is not demonstrable let alone observable. But it <u>does</u> dichotomize your awareness, and that is demonstrable as Gazzaniga showed. **Please understand that I am only using the words** *consciousness* **and** *awareness* **in the way I define them**.

In the optical device of this imagined experiment, we can substitute the kitten with Prof. Schrödinger's view of a garden and substitute the puppy with a view of a darkened room. Interchanging them produces no scientifically demonstrable difference except one of parity – the two halves of your body and brain are mirror images of one another, so are not exactly equivalent for the purposed of the experiment, but the parity difference does not undermine the conclusion.

A simpler experiment is the following: As you look straight ahead of you right now, your right and left visual fields arise in two totally different configurations of atoms in your eyes, optic nerves and brain hemispheres, but there is no basis for attributing your conscious experience of what you see to two different conscious selves. Such a distinction is factious. This remains true if you turn off your corpus callosum. You stare at the vertical bar in this table without peeking at the letters:

C	|	A
Z	|	G
N	|	H
Q	|	N
H	|	Q
G	|	Z

You are asked to point at the letter in the right column that corresponds to the one in **bold** print in the left, you can do so with either hand if your corpus callosum is turned on, but with neither hand if it is turned off because a signal across the corpus callosum between the right and left hemisphere is missing. A person with an interdicted corpus callosum will quickly learn to solve this problem by putting both letters in the same visual field, first the left then the right after which they can point with either hand. This suggests that having a severed corpus callosum did not seem to be a problem, for this sort of reason, until Prof. Gazzaniga came along.

You in consciousness and you in awareness

Since you are not aware of the experiences of others, you do not share their memories, pain, joy and so on, except in compassion or empathy, it may seem that there is no point in talking about the universality of consciousness. Why not just dice it up along the lines of awareness. We do not live in an age of organ transplants that include the brain, but someday may. Suppose you were blinded by meningitis destroying your visual cortex, and I was killed in a motorcycle accident but donated my brain to the brain bank and it was found compatible with yours. After your damaged tissue was replaced by mine, you can see normally again. Assuming no cybernetic incompatibility, there is no need to perpetuate the division of consciousness along the lines of the boundaries of awareness between you and me. The "you" that can see again and the "you" that I was are the same "you" in consciousness. *Tat tvam asi.* Only the binding in awareness of my sometime brain tissue has changed.

The Samkhya paradox and you

The Samkhya paradox appears to each of us in a more fundamental way: To keep the awareness of consciousness in the forefront please allow me to use you as an example. You are aware of your own consciousness but not aware of the consciousness of anyone else alive in the sense that you do not say "ouch" when they hurt. Nor are you aware of the consciousness of anyone who lived before you were born nor anyone who will live after you die. What is special about you in consciousness in all of human history? You're only one of billions of people and you are made of atoms that were selected by the happenstance of where your mother did her market shopping. Bob Dylan may have been singing about your awareness when he sung "Been a long time comin' and I'll be a long time gone".

It is easy to understand why you are not aware of being here a century ago and why you won't be aware of being here a century hence, but why did billions of years go by before you became a conscious self and then your conscious self-hood will disappear for billions more. This is the Samkhya paradox. There is a simple way of getting rid of it: Get rid of the delusion that creates it and that can only be done if you make the distinction between consciousness and awareness made here. There is only one conscious self: You. The Vedanta principle that you are the conscious self may better be expressed by saying that conscious self-hood is not denumerable. If anyone wakes up in the morning, or ever woke up in the past, or ever will wake up in the future, in consciousness it is always you. That's why you woke up this morning. It had to be you. In awareness however you are unique: You cannot remember being here 100 years ago won't remember who you are now 100 years hence. In awareness you are demonstrably distinguishable from everyone else today, and probably everyone else who ever did or ever will exist. By the same token you are not aware of being anyone else. Your conscious states do not bind with theirs.

There is no scientific reason to believe that our personal awareness such as memories will survive the crematorium furnace, unless memories, like breast sucking, came to you via your genetic code like the ancestral memory of how to build the nests of Schrödinger's birds, or were immortalized in images like photographs and books and word of mouth of your mother tongue. Science has nothing to say about God remembering your escapades nor allowing you to forget them. Hence science cannot affirm or deny such belief in reincarnation. Scientifically, your memories which are states of awareness are erased by death. Beyond personal experiences, most of what we have in our awareness that came from others before us is language, knowledge, tradition, intelligence and emotion. These are immortal, given the survival of the human race. However, in the above assessment of the relation between awareness and consciousness, self-hood in awareness and self-hood in consciousness, if a brain similar to yours in the future experiences the same conscious state as you did, both of these experiences happen to the same conscious self. Your consciousness experience of hearing Mozart's music is, in part, a reincarnation of some part of Mozart in you.

The Vedanta principle: **Tat tvam asi** implies universality of consciousness, but not universality of awareness as Gazzaniga has proven. If a human body and brain substantially equivalent to yours exists in the future beyond your life one may say that both have the same conscious self. That should not be construed to mean that both share the similar patterns of neural firings. A cybernetic analogy would be to equate your body and brain to a computer and equating the program, a pattern of bits stored in the computer memory to your awareness. Evil spirits of biblical times may be seen as unfortunate patterns of neural firings. The mission of theistic religion if not all religion is to get rid of the bad spirits and embrace good spirits. The agnostic does not disagree but claims to need no avatar. There may be little anatomic difference between the Dali Lama and a criminally psychotic ISIL recruit, but there is a big difference between the patterns of neural firings in their brains.

The binding of consciousness thought "experiment"

There is perhaps a more stark way of describing **Tat Tvam Asi**: Suppose you suffered from almost total amnesia after being struck by lightening, but did not loose basic skills like language, the ability to read and write, or the basic skills of body care. But you recognized no friends or relatives and remembered no special skills. Going home from hospital and meeting friends and relatives and looking at your house and home and family photo albums was just like going into any house of total strangers and looking at their family, their home, their photo albums.

This is clearly a scientifically demonstrable dichotomy of awareness in the sense I have defined that word. But there is no scientific way of demonstrating that you are a different conscious self. You are the same conscious self, grateful you can still remember how to speak your mother tongue notwithstanding other damage in awareness. Knowing your plight, you choose to start a new life with a bunch of complete strangers in a strange house, because they all claim to recognize you and love you. Why not? No other family does. You enter a new state of awareness, to start a new life, with no recollection of your past personal memories. Suddenly one day, perhaps a year later, you recover your memory. The memories of your former life and present life rebind in a single continuous state of awareness. There still is only one conscious self. There never were two. *Tat tvam asi.* The two disparate selves that became one were states of awareness, not consciousness.

Consciousness binding experiments you do not need

I read every popular article I find on brain function but have never read one that explains the binding of consciousness in terms of neural activity. The only brain I can turn to is my own. I would not pester you with personal experiences but I have learned something from my own defective or unfortunate brain activity, described below.

The following "experiments" further clarify the distinction to me between consciousness and awareness and why I believe that awareness, as I have defined it, is the binding of consciousness, as I have defined it. I would not wish them on anyone. Two decades ago I woke up and my right ear was ringing. I later found that it rang at 13.7khz. It rings all day and all night, week in and week out, year in and year out. There is positive feedback in the inner ear amplifiers to increase gain but mechanisms exist to prevent oscillation, one of which went kaput. Neurologists explained the tinnitus problem to me but could offer no help. I was on the verge of suicide, but as the months wore on I ceased to notice the ringing – unless reminded of it. I do not believe it stops even when I am asleep. Now, days can go by without my being aware of it – unless I think about it or I am otherwise reminded. Then I am aware that it never has stopped ringing. Of course, I can hear the ringing now because I am thinking about it. I cannot think about it without hearing it.

The second "experiment" happened to everyone where I grew up on a gold mine. We lived about a mile from the reduction works where rock crushers – both stamp mills and ball mills ran "24/7" except on

Christmas day and Good Friday. Even with all the doors and windows of the house shut, it sounded like the ocean – but constant: not rising and falling with the waves. We all learned never to notice it, even though the sound was vibrating our eardrums continuously day and night. It was only starkly apparent twice a year. On Good Friday morning and Christmas morning when one awoke to the sound of silence. I can never listen to the Simon and Garfunkel song without remembering what it felt like.

I don't believe that not noticing these sounds turns off the consciousness associated with the pattern of neural firings that causes them and which never stops. The conscious self always hears the sound. Whoever I am in awareness turns off the equivalent of the corpus callosum that binds my self awareness to the noise, unless attention forces me to be aware of it. Thus awareness, which is scientifically demonstrable, is the binding of conscious states that are not demonstrable, because even though they are real to me, I cannot demonstrate their reality to you. I have no cause to believe that the tinnitus part of my brain that oscillates day and night is aware that is is conscious let alone aware of its own existence.

Presumably science can discover the relationship between objectively describable neural firing patterns and the awareness of consciousness. Before doing a partial visual cortex brain transplant, the neurosurgeon would have to make sure the neural firing patterns match between your well brain tissue and the tissue being replaced by mine, for example with the perception of color. This is a scientific job that can be done by a colorblind neurosurgeon who does not know nor need to know what it feels like to be either of us.

The non-demonstrability of consciousness means that being aware of the pattern of neural firings associated with some conscious state does not convey what it feels like, unless both the awareness and consciousness is happening to you.

How can you be aware that you are conscious? Leibniz was. Yet most people pass the whole day without being aware they are conscious but see themselves as being in a physical world, taking consciousness for granted. Yet when they close their eyes and listen to music, they are indulging in consciousness and that alone especially if the music is not representational as in a song or dance. One is aware of consciousness if one wants to get back to listening to music after having to answer the phone. When you dance it may be your cerebellum that is having the most fun: It is more your partner than you.

Consciousness in antiquity

The ancients believed consciousness to be deceptive and unreliable. Example: Put one hand in a tub of water as cold as you can stand and the other in a tub as hot as you can stand. When you can no longer stand either put then both in a tub of lukewarm water. It will feel hot to the hand that was in cold and cold to the hand that was in hot water. But the water is only one temperature but one hand says *hot* the other says *cold*. The ancients did not understand that the senses re-calibrate to give measurement on a relative scale. On a bright day we see white light on a scale from the darkest shadow as black to the lightest high light as white. In moonlight which is a million times less bright than daylight, we see the same thing but the lightest white in moonlight is 100 to 1000 times fainter than the darkest shadow in daylight. It is the relative scale that conveys intelligence. Morse code is intelligible if the dashes are relatively longer than the dots. If not, making them louder or both longer or both shorter does not help.

Descartes

Descartes' *Cogito ergo sum* (I know therefore I am), says too much. Feeling the sun on a sunny day is a real conscious experience for all, including forty percent of Americans who believe the sun orbits the earth. Instead they should say: *Cogito ergo non compos mentis*. They vote, which could account for at least forty percent of the government. What they "claim to know" which is what they believe and what they feel is real should not be confused.

We do not choose to believe consciousness is real. We experience its reality without choice.

Vehemently believing consciousness is not real is a real conscious experience. Descartes explains, *"We cannot doubt of our existence while we doubt ... ."* But be careful about his word *existence.* The meaning of the word can change with the person using it. Doubting is a real conscious experience but consciousness does not exist in the sense that the physical universe is said to. Consciousness does not play second fiddle to belief. If the conscious self experiences consciousness, who is the conscious self: You? I? Him? Her? God? Instead of saying *I know therefore I am,* one really has no cause to say other than: "An experience in consciousness of being amused at your joke is real. Since there are only two of us and you told the joke, I infer that I am the one amused". One can forgive Descartes for saying *I know therefore I am.*

Consciousness in the modern world

What we experience consciously is geared to the relative scale of awareness. The objection of the ancients pertains to awareness because you could use someone else to measure the lukewarm water temperature and be aware of the same results without experiencing consciousness yourself. If you made a machine that does the same relative scale measurement, there is no need to make any assumption about whatever states of consciousness may attend its operation. Paint stores have machines that accurately match paint color but there is no reason to believe the machine experiences similar consciousness to a human seeing the colors. The converse is also true.

For survival purposes, we have no need of senses or conceptions that account for the physical order beyond the approximations that the velocity of light is infinite and Planck's constant is zero. At least we did not until the end of the 19th century. Excrement and rotten food stink. The repulsive stench is associated with a pattern of neural firings that evolution has chosen to map reality to dissuade higher life forms from ingesting pathogenic bacteria. Lower life forms like flies and dung beetles not so threatened no doubt find the aroma of rotten meat delectable. Blue cheese is rotten and tastes good even to us.

When we are aware of anything we consciously experience something which is associated with a pattern of neural firings in the brain. Whenever that same pattern occurs, we have the same conscious experience, say seeing blue. There is no reason to believe that a particular pattern is the only one possible to represent blue. It is simply the way blue 3500Å photons entering your eye are mapped in your brain. If you suffered a brain injury that blinded you in part of your visual field; and I killed myself playing with my new AK47, but had donated my brain to the organ bank and it was used to fix your the damaged areas of your brain, you may tell your neurosurgeon that you can see again but the color of the sky is wrong where your vision was restored. Q. "What color are you seeing?" A. "I do not know – it is a color I have never seen before. Evidently the color my donor called *blue*". Evolution may have more than one good way of solving the seeing-blue-problem. One would hardly expect that every time blue light entered your visual field where you were once blind, it made your big toe feel cold, or you would have to believe they my seeing the blue sky felt like cold toes.

In a technical dictionary I saw acetone defined as a clear organic liquid solvent with a pleasant fruity aroma. To me, the smell of acetone is nothing of the sort. It stinks. I have a cousin who since childhood never ate fruit. If it tastes to him like acetone smells to me I wouldn't touch fruit with a barge pole either. One may acquire a taste for beer or wine but I will never acquire one for acetone. I can only account for this difference by supposing that the genetic codes have different ways of mapping in consciousness the smell of the same substance. After changing my engine oil, the mechanic at the oil change depot told me to stop using Pennzoil. "It is bad for your engine!" he replied. I asked him how he knew that. He dipped my dipstick in Pennzoil and invited me to smell it. "It smells bad, doesn't it?!" he said. I agreed, but told him I'd go on using it because my engine loves the smell of Pennzoil, so it must be good for my engine. To his credit, he seemed skeptical.

Consciousness as we experience it is associated with neural firing patterns in our central nervous system. In the heyday of high fidelity audio, a symphony was composed for orchestra and tape recorder. Portions of the performance were prerecorded. The audio equipment used was off the shelf but top of the line. The audience was flagged to close their eyes and guess what they were listening to; the tape recorder and loudspeakers or the live performance with the loudspeakers quiet. It was not possible to tell the difference except in loud passages amplifier distortion was evident. Both vibrate the eardrums in the same way so consciousness is downstream in the chain of causality. Similar reasoning applies to vision. Yo-Yo Ma is not a tape recorder and his cello is not a loudspeaker. This establishes our consciousness as being associated with neural activity in our brains and not directly with the atoms of cellos or loudspeakers. It does not give us any reason that it is or is not associated directly with any other activity. There is no reason to believe that consciousness is anchored to a point in time but in the brain there is a reason to achieve sufficiently high time resolution commensurate with muscular activity that may be needed for protection against danger.

If we talk about consciousness, it is clearly important to be aware that we are conscious. For most of the day most people are not. When they enjoy instrumental music it is the conscious experience they enjoy even if they do not think about consciousness. The Humpback whale probably does the same thing while singing a twenty minute aria. We assume that ultimate reality is the external physical world, ignoring the fact that were it not for consciousness there would be no reality to our experience. Our existences would only be an abstraction. The whale may be able to distinguish the conscious experience of a composed aria from that of awareness of physical surroundings. Could Descartes imagine awareness of consciousness on the part of a being without self awareness? How can "I know therefore I am" if I do not know I could be. An electronic oscillator may experience consciousness but certainly does not know that it does, much less does it care whether it even exists. A small child with a ringing ear may think everyone can hear it. She could not learn that only she is hearing it without self awareness. Being consciousness and self aware does not guarantee awareness of consciousness; its absence for anyone does not imply the non-reality of their consciousness.

These facts about our conscious experience may seem to be a compelling argument that there is no reason to believe that consciousness is anything but some spin off of higher brain function: An argument that wins debating points from those looking for confirmation of their preconceived beliefs. Those looking for the truth need to counter it with this observation: Higher brain function is always presented to us in consciousness and never in any other reality **because there is no other reality.**

Higher brain function is an incredibly complex and sophisticated construct of awareness that goes beyond scientific analysis of the neural firing patterns of cognition and their relationship to behavior. It is locked to subjective consciousness that is beyond the reach of science to understand, but an essential component of life. Science can understand that those who taste sugar report experiencing the taste of sweetness and may even be able to record the associated pattern of neural firings, but science cannot observe that conscious experience or understand why it is real. It is a miracle.

Phantom limbs

It has been seen as a paradox of consciousness that it is very common for amputees to go on experiencing conscious sensations in an amputated limb: "How can I have a pain in my right leg if I do not have a right leg?" This paradox vanishes in the mind of scientifically aware people in the past century with the understanding that everything that is consciously real to us, every moment of every day from birth to death, is a construct of our brain. The brain can construct a consciously experienced model of space and locate sensation in it with high accuracy. Pedestrians step in front of your car as you approach a pedestrian crosswalk. They bet their lives against the desire to get to the other side of the street that the model of reality in your brain is accurate. Although present day prosthesis do not have sensory and neuromuscular feed back, if you a walking with an artificial leg there is nothing phony about feeling both legs functioning properly. Your prosthetic nurse knows this: If you ask: "How can I feel where my right leg is if I do not have a right leg?". She can honestly answer your question: "You do now: It may not have a feedback loop to the brain, but it really helps you walk the way you feel you do. The model of reality is in your brain, not your leg."

The Honey bee

There are about one million neurons in a bee's brain and there are a billion synapses. Honey bees are smarter actually than lobsters. Bees and butterflies evolved in a symbiotic relationships with plants, resulting in flowers. The yantra of petal forms attracts the bee as do the colors of the petals. The aroma draws the bee closer and finally the delectable flavor of the nectar is the bee's reward for pollinating the flower. Flowers evolved to satisfy the conscious experience of bees long before we were here. Yet we, with one hundred thousand times as many neurons and one million as many synapses are awestruck by the beauty of flowers and softness of their petals the cultivation of which supports large industry; exotic perfumes commanding high prices are made from flower oils to satisfy luxury markets; we enjoy the diverse flavors of the floral nectars in honey. The bees and the flowers did all this and didn't ask us for our advice. Thank God. Had they done so, they would not exist. In the cultivation of flowers we are late comers to their evolution. We cannot claim consciousness as the property of large brains. In awareness, the brain of a bee could be simulated with a microcircuit of ten million transistors. One bee is not very intelligent, but there are tens of trillions of them.

Baboons and bees may be incapable of being aware that they are conscious, but flowers to me are the proof that bees _are_ conscious, not because they are attracted to flowers which is scientifically demonstrable while their consciousness is not, but because flowers are consciously attractive to _me_. I cannot prove scientifically that I am consciously attracted to flowers, but to be conscious of them, I do not have to.

Do computers experience consciousness?

On my desk there is a 20 year old Pentium II computer that I keep off-line to prevent virus infection. I still use it to run several programs. A machine of the same type as mine loaded with an Intel chess playing program won a game of chess against the world chess champion Garry Kasparov in 1994. The Pentium II is a single crystal of silicon about the size of a postage stamp doped in selected places with two other chemical elements. It is metaphysics to proclaim that Kasparov's brain, playing chess, experiences consciousness but the Pentium II does not. It is certainly true that the Pentium II was not aware that it was conscious, nor even aware of its own existence, or even that it was playing chess, or even that it gave a tinker's cuss whether it was winning or loosing. It was not programmed for these tasks. It would be very interesting to figure out how to do so, but that is another story.

If one asks whether a computer experiences consciousness akin to that experienced in association with human awareness, I would hazard the guess that a Pentium II playing chess experiences a rapid staccato of conscious states with a processor instruction repetition rate measured in nanoseconds. They do not bind in an overview of the game which lies in its analog of Freudian sub-consciousness: dormant instructions stored in memory. By contrast, Kasparov has an overview of the game by virtue of the massive parallel processing of the brain: Although his brain goes through a rapid sequence of consciously experienced thoughts, recent past thoughts relating to the current move are available for support from short term memory. He is emotionally motivated to analyze the current move in the light of its importance and has a broader emotional motivation towards victory. In the light of *tat tvam asi* the emotion motivating the Pentium II game is in the brain of the computer programmer: The Pentium processor and its memory, like his own cerebellum, lies in his id. There is an amazing story of a grand-master who lost a game to Bobby Fisher. While analyzing the game later to discover how he lost a game he believed he should have won, he discovered that Fisher won because, in say move 20, Fisher made an obscure but ingenious move. When he asked Fisher how he thought of it, Fisher replied: "I didn't think of it. I remembered it was a move made in a similar situation described by say, Capablanca in his, say, 1927 book". I have used the word *say* to cover my bad recall: unlike Fisher's who could remember the important points of every game he ever won or lost or anyone else ever won or lost that Fisher had read about.

The reality of consciousness is not demonstrable whether or not it happens to a Pentium II, Kasparov or to me. <u>It demonstrates its reality only to you.</u> You cannot demonstrate it to anyone else. As far as chess goes, it has been proven that a Pentium II can be made **aware** how to play chess: It passed the Turing Imitation test. If it is conscious, it will never be able to prove it to you, unless of course you are a Pentium II too. Even then it would take some sort of microcircuit transplant in which case the awareness of my Pentium II and you would unite. There would still only be you. *Tat tvam asi.* There would still remain the problem of programming the resulting machine to be aware that it is conscious.

If we ignore bees' consciousness creating flowers and computers defeating the world champions in mind games, it is very easy to come to the conclusion that consciousness is only a property only of higher brain function. That it is some kind of phenomenon or substance. That one day, perhaps, we will have instruments that will enable us to observe it and study its properties: <u>This view is a fatal trap.</u> To avoid falling into it, and remembering the caveat about not confusing consciousness and awareness as I define them in this paper, I refer you to the present views of Peter Russell (You Tube) and I restate again Erwin Schrödinger's ageless view:

Consciousness cannot be accounted for in physical terms
For consciousness is absolutely fundamental
It cannot be accounted for in terms of anything else

We cannot presently say if consciousness as we experience is only associated with neural firing patterns or with atoms: electrons, protons, neutrons and photons of the material that constitutes our nervous system. But this question is in principle answerable by partially replacing brain tissue with electronic circuits of different atomic composition. But some part of our brain must be aware of both conscious states. This is certainly possible because we can demonstrably distinguish color from musical pitch from smell etc.

Do electric motors, for example, experience consciousness? The important thing to understand about this question, for the purpose of this epistle, is that it does not have to be answered. The only consciousness we can account for is our own. I believe you are conscious but, short of brain transplants, I can adopt a solipsistic view. The same is true of electric motors. I have no reason to believe that consciousness of electric motors is not real, but I am sure the following is true of them:

- They are not aware they are conscious
- They are not aware of their own existence
- They give not a toot even if the bearings are seized and the windings smoke
- If I could be an electric motor for a day and return to being me, I would remember nothing of it
- There is only one conscious self. I don't have to be an electric motor because I already am
- I am not aware of being an electric motor because there is no

corpus callosum connecting electric motor wiring to my brain

- There is no present cause to establish a Society for the Prevention of Cruelty to Electric Motors. But, one day, when we much better understand the relationship between sensation, neural activity and current flowing through wires, we may learn that single phase motors feel pain and do not like it and three phase motors feel pleasure and love it. Or something. Then there oughta be a law against single phase motors. (Probably not a bad idea anyway).

Human consciousness is associated with activity of the cortex of a brain, an enormously complex living organ. It may seem absurd to talk, in the same breath, of this consciousness and the activity of a photon in an interferometer. The *Physical Universe* seems very real to us because:

- Plank's constant is so small:

  $h = 0.00000000000000000000000000000000006626$ Js

- The speed of light is so large: $c = 300,000,000,000$ mm/s.

It is reasonable to describe brain function in Joules, J, and millimeters, mm and seconds, s. The approximations $h=0; c=\infty$ are very accurate in these units, so the *Physical Universe* is a very accurate approximation to reality: <u>In these units.</u> But a miss is as bad as a mile. In fundamental particle physics, different units are used which make $\hbar=c=1$. Now the *Physical Universe*, in these units, is so bad that only Judith Viorst can give us the words to describe it: as absurdly worse than an extremely terrible horrible no-good very bad approximation. But consciousness is not a metaphysical belief or an approximation. It is just as real no matter what units of physics we may don as spectacles to view reality in awareness.

Redefining units to make $\hbar = c = 1$ makes for a great simplification in understanding the physics of the interactions of elementary particles. If you are a non-physicist who is not terrified by formulas, mostly simple, and have not already done so, I recommend viewing Prof. Leonard Susskind's online lecture: Super-symmetry & *Grand Unification: Lecture 1*. Like Prof. Feynman, he has the ability to make the impossibly complex sound easy without conceding to analogy and metaphorical hand-waving. After that, there are dozens more of his online lectures on Stanford University physics continuing education lectures. He transcends even Feynman at getting the ideas of physics across. He assumes some math knowledge but no more than is covered in the appendices of this screed.

The brain is not constructed to make intuitively obvious solutions to quantum mechanical problems, but to solve problems in the "physical world" like looking for our lost reading glasses and judging at our current speed whether it is safe to try and stop before the light turns red without quantum "uncertainty". It is designed to find single deterministic solutions. Quantum "uncertainty" may happen in the brain but it should not lead to uncertainty about what the brain predicts will happen in a deterministic world. We are aware of what is called "quantum uncertainty" in the laboratory because the brain is on the receiving end of an experiment which requires the quantum mechanics to be understood.

We have great veneration for the physical world because there is a lot more of it than the atoms in our skull which have to map all
10000000000000000000000000000000000000000000000000000kg
of the universe with a brain weighing 1 kilogram. We should not be hard on the brain for not getting it right all the time.

Our brain is wired to make us feel ourselves to be inside a model consistent with complex observational evidence, not only of out direct senses but with sophisticated extensions of them with the likes of forensics and the Hubble Space telescope. We can improve the model with imagination. Until the late 19th century there was no evident reason to believe that we could not in turn create absolute reality in the image of the model. Over 2,500 years, the annals of science accumulated the theoretical elaboration of the model and the enormous body of evidence that supported it. Then suddenly in 1905 and for a few following years, the metrological trinity of its foundation since antiquity, mass, length and time were shown to not to be absolute. A decade later its geometry was shown to be approximate and a decade after that the theoretical core was shown to be impossible.

The average person does not feel anchored to these scientific things, but they are anchored to it in another way: It is hard wired into their nervous system by evolution, or God if you please, and presented to them in intuition. We feel ourselves to be in the Physical World. Its stuff is what it seems we are made of. The currently understood laws of physics revert to this model if we make two approximations: the velocity of light is infinite; Planck's constant is zero. For the average person the approximate model is very accurate in awareness. We have never been lied to by experience: <u>The model has never been presented to us in reality as anything but consciousness.</u> The same is <u>not</u> true of awareness: We can be aware, for example, that the earth is flat because plumb lines are parallel, as near as we can tell, with our hardware store level.

Although intuition about physical reality is not a true model of quantum reality, there are two respects in which intuition tells us truth that can only properly be understood by quantum mechanics. These are discussed in the chapter: **Arrow of Time**.

Consciousness not made of the stuff of the Physical World, as von Neumann imagined. It is made of Absolute Nothingness. The structure of its states is the quantum mechanics. If you think electrons are "in the physical world" try catching one. As soon as you know where it is you cannot predict where it will be next. But if you know how it got somewhere else, then it can be anywhere. Then if you have two of them you cannot tell which is which. It is pity Shakespeare didn't know about them – he would have had fun with them in *A Midsummer Nights Dream*. If Mozart had known about them, the villain in the Magic Flute may have sung: *"Der Elektronfänger bin ich ja ..."*

Hugh and John on psycho-physical parallelism

Both Hugh Everett and John von Neumann recognized that this doctrine cannot be ignored.

Everett: *"...to assume that all mechanical apparatus obey the usual laws, but they are somehow not valid for living observers, does violence to the so-called principle of psycho-physical parallelism, and constitutes a view to be avoided if possible.* Everett keeps the door open to the quantum mechanics representing consciousness but does not close it as representing physical reality.

Von Neumann said: *"First, it is inherently entirely correct that the measurement for the related process of the subjective perception is a new entity relative to the physical environment and is not reducible to the latter. Indeed, subjective perception leads us into the intellectual inner life of the individual, which is extra-observational by its very nature (since it is taken for granted by any conceivable observation or experiment). Nevertheless, it is a fundamental requirement of the scientific viewpoint—the so-called principle of the psycho-physical parallelism—that it must be possible to describe the extra-physical process of the subjective perception as if it were in reality in the physical world—i.e., to assign to the parts equivalent physical processes in the objective environment, in ordinary space.*

Here von Neumann asserts that the quantum mechanics represents consciousness, but seals in the Physical World in ordinary space: the objective environment seen as reality.

I believe that von Neumann should have put it this way: *Nevertheless, it is a fundamental requirement of the scientific viewpoint—the so-called principle of the psycho-physical parallelism—that it must be possible to describe the extra-physical process of the subjective perception as reality. The physical parallel is the experience of a model in awareness that may be approximate or even fantasy. The objective environment is mathematical, with approximations yielding physical world models.*

The compelling objection to von Neumann's *subjective perception as if it were in reality in the physical world* is that the "objective environment, in ordinary space" could not be represented by the quantum mechanics without one being forced to bolster this metaphysics with further metaphysical attributes in which causality is broken and predictability random. EPR used the term: "Objective reality" which like von Neumann's "objective environment, in ordinary space" defaults to "The Physical World".

Both Everett and von Neumann realized that the quantum mechanics must be, at least, a descriptor of consciousness. The thrust of this epistle is to make the point that the paradoxes of the quantum mechanics or the need to create multiple physical universes disappear when belief in the reality of the physical world of classical physics is abandoned, beyond seeing it as a mathematical device and always an approximation to some other. This leaves no reality but consciousness, which evidently does not exist, but remains absolutely real in a way that does not depend on what we belief, but is beyond belief.

Gazzaniga returns

In the light of the above, we return to a Gazzaniga inspired Gedankenexperiment to clarify the distinction between conscious self-hood and self-hood in awareness in brain function. Were the following experiment actually possible it would probably be considered unethical if not downright immoral and illegal. But it is only a thought experiment so we brush these considerations aside. Skip it if it makes you queasy.

You and I share the same hospital ward after each has had a slice taken out of our corpus callosums to install remotely controllable electronic interfaces. The neurosurgeon has installed two temporary cable connector sockets on the top of each of our heads, each with one hundred and fifty million pins, used to allow cable connections to a computer that programs the interface to make sure it correctly matches corresponding nerve endings of the interdicted corpus callosum. Assume that this has been done so each pin on one connector connected to the corresponding pin on the other correctly makes a correct connection between surgically severed nerve endings. Prof. Gazzaniga has been doing some interesting experiments on us, but right now he and our neurosurgeon are playing a round of golf and have left us with jumper cables that reconnect our corpus callosums and instructions to pull out the jumpers if we feel an epileptic seizure coming on. This would electronically re-sever our corpus callosums.

They forgot to lock the cable storage box so we decide to do an experiment of our own while they are away. I plug a cable into my right brain corpus callosum and you plug the other end of the cable into your left brain. You likewise plug a cable from your right brain to my left. We ask another patient on the ward, in for a knee replacement, to help us. He hobbles over and pokes your right side with a pencil after we close our eyes. The game is "With your right hand, point to where you were poked". Without hesitation you point with your right hand to the spot you were poked. We close our eyes again and ask our assistant to poke a different point on your right side. The game now is "With your left hand, point to where you were poked". I immediately reach over with my left hand and point to a spot on my body that corresponds to the spot on your body where you were poked. My brain only has a map of my body and it cannot distinguish the signal it receives from one it would have gotten from my other brain hemisphere which sent no signal at all.

The purpose of this Gedankenexperiment is to ask two questions:

- Who is the self in awareness who knows where he was touched
- Who is the self in consciousness who feels where he was touched

The first is a scientifically answerable question: A being made of half of you and half of me can identify where your body was touched by pointing at the corresponding spot on my body which serves as a map. The other self in awareness made up of the other halves of our bodies and brains experienced nothing and is unaware of being touched (unless I actually touch the spot on my body where I feel you were touched): *Tat Tvam Asi* does not come to the rescue, as Prof. Gazzaniga proved.

The second question is not scientifically answerable but presents no paradox if we see the conscious experience happening to a unitary conscious self who feels the touch and can point to a point on a map of where the sensation was felt. *Tat tvam asi.*

Consciousness - Awareness; Reality - Existence

In the light of the preceding, the purpose of this chapter is to attach unambiguous meaning to the four words of its title. If, after you have read the following and find yourself in semantic disagreement, please search and replace my words with the words you think I should have used instead. I do not want to diminish the importance of semantics, however interesting it may be in contexts unrelated to the purpose of this paper, but to make the point that whatever semantic choices one makes do not change the conclusions.

The twentieth century brought events that not only shattered the ancient idol: The Physical World: It undermined the very premise of scientific theory: the relationships between pointer readings. The resolution was the quantum mechanics that emerged over the first three decades of the twentieth century. It is the most successful and far reaching theory of scientific history, but the metaphysical thunder of its apparition continues to this day.

It may appear that associating consciousness with the quantum mechanics would require a scientific investigation of consciousness, treating it as a phenomenon, revealing its structure, before we can begin talking about physics. Then we could see how the structure of consciousness fits into the quantum mechanics. After all, to find out how gasses behaved in the light of Newtonian physics we had to relate gas pressure and temperature to the dynamics of gas molecules as a first step to developing thermodynamics.

The above view, although apparently natural and reasonable, only embeds one more deeply into illusory classical metaphysics. I repeat the paraphrase of Condon and Shortley's words: Classical physics formulates the laws of physics as relationships between the observed data, like pointer-readings, given by observing apparatus. All exact science has proceeded along such lines hitherto. **Quantum mechanics does not do this.**

- It is impossible to perform a scientific experiment on consciousness because it does not exist: We cannot observe it.
- We only experience it
- We cannot infer its structure by introspection
- We cannot experience its subjective content given only objective descriptions of brain function we know are associated with it
- The above premise debars us from ever proving the reality of

consciousness beyond being that proof ourselves
- We are forced to accept the reality of consciousness as miracle

The proof of the reality of consciousness is given to you, but you cannot prove it scientifically to anyone else without blurring the distinction between your self-hood in awareness and theirs. However, its reality is not inconsistent with science. What we are asserting in this epistle is that the quantum mechanics, as best we understand it, is the basis of awareness, the structure of consciousness without which nothing would be real and awareness only an abstraction. This is seen to be equally true of the physical world, ergo the doctrine of psycho-physical parallelism. Most physicists today still hang onto the Physical World, keeping it alive with Copenhagen band-aids: The belief that it is ultimate reality is a convenient illusion.

The human brain and presumably the brains of higher animals model the universe in awareness, but reality is never presented to us directly except in consciousness: We consciously experience being aware of something. We can be aware of being conscious, but usually are not, particularly involving events that do not involve the atoms of our own bodies and brain. For example, approaching an intersection the light turns yellow. We are aware that it will soon turn red and we want to stop before that to avoid death and destruction that plays out in awareness as events in the "external physical world". But the whole screenplay of awareness is projected on the screen of consciousness of which we can be aware, but usually are not. Our brains are not solving the equations of quantum mechanics or relativity: Our central nervous system is using a much simpler but very accurate model and that model is calling the shots to predict what will happen in the $h=0$ and $c=\infty$ "external physical world" which we assume is "ultimate reality", without stopping to think that it never is manifest except in consciousness.

When the quantum mechanics arrives, Heisenberg gives no thought to consciousness as reality. Von Neumann does, but cannot abandon the belief that the physical world is ultimate reality, and sides with Heisenberg and Dirac presenting a view of reality that is intrinsically random, with observation instantaneously and everywhere destroying unobserved reality. To escape from this bizarre metaphysics, Einstein proposes an add-on that it is a regression to the "physical world", later shown to be impossible.

More than a quarter of a century later, a year after Einstein's death, a physics student at Princeton showed that the truth has been lurking unseen in the mathematics of the quantum mechanics: Reality is infinitely more vast than awareness can possibly imagine. This limitation is not a practical limitation on awareness like "all the disks are full" or "there isn't enough paper and ink in the universe", but lies at a very fundamental level of physics. The first clue was hidden in Schrödinger's Nobel prize winning paper *"Quantization as an Eigenvalue problem"*. Everett, the Princeton student, showed that states of awareness are caused by the eigenstates of the observation operator, restoring causality to physical theory after von Neumann abandoned it. Everett did not commit to an interpretation of reality as we are doing here. Identifying reality as consciousness gives us a very simple and elegant view of the quantum mechanics: The states of consciousness that correspond to each state of awareness does not bind in a single state of awareness, because each is caused by an eigenstate of the operator that represents the observation we have made, and all of these eigenstates are mutually orthogonal. If we are aware of being in one of the eigenstates we cannot be aware of being in any other.

Heisenberg and company assumed that we will be aware automatically of everything that is real. We protect this delusion with the metaphysics that the vast reality of which are not aware, which must be preserved to explain interference of amplitudes, must be "collapsed" into non-reality at the moment of, and by, our act of observation. Any of the collapsed states could have been real and will be so from time to time if we repeat the same experiment many times. These mental contortions to preserve Physical World "reality" force us to the belief that the only reality we do experience is chosen at random. What Heisenberg did is absolutely necessary if one believes that the Physical World is reality and if one rejects hidden variable theories.

To this day most people accept "the physical world" as being ultimate reality. Everett's Relative States Formulation of the quantum mechanics cannot, without absurdity, be reconciled with this belief. This is not true of consciousness, because two contradictory states of consciousness can both be real as long as they do not bind in the same state of awareness. This will be true if they are caused by orthogonal eigenstates of the observation.

The bizarre paradoxes of quantum mechanics come from the belief that it describes reality not as consciousness but as "the Physical World". Electrons and protons, photons and gravitons are not "physical objects" but mathematical entities that underlie the structure of consciousness. They do not obey the laws of intuition and classical physics. The quantum mechanics does not explain the subjective content of consciousness, such as the taste of sugar, even if we can scientifically relate the subjects <u>awareness</u> of sweetness to known patterns of neural firings. For example a subject, given cubes of table salt, Epsom salt, sugar and chalk that all look the same puts them on their tongue, correctly identifying the sugar cube as sugar thereby demonstrating scientifically to you the scientist that they are <u>aware</u> which was sugar. There is no scientific demonstration of what it feels like to taste sugar, or anything else, unless you are subject; then it is science only between God and you.

The sense in which I am defining the words below does not require understanding of, nor depend on, the way others may use them.

Consciousness

It is impossible to define consciousness for the reason given by Schrödinger: It cannot be accounted for in terms of anything else. It cannot be the object of a scientific experiment: It does not exist. It, is experienced, like a bad headache. I chose this example rather equally valid conscious experience like orgasm or hearing a bird sing, which could be construed as observably coming from someone or something else.

I cannot demonstrate headache consciousness to you by handing you a flask of headache consciousness. It does not exist. I can demonstrate the reality of consciousness just by putting my hand on a hot stove, but it is no use asking you to tell me what it felt like because you are not *aware* of what happened, there being no corpus callosum between your brain and mine. Your states of consciousness and mine do not bind in common awareness. If your corpus callosum has been replaced by an electronic interface and you put each of your hands on a hot stove, the pain you feel in each hand will be experienced by the same conscious self. If someone turns off your corpus callosum while this is happening, the pain you feel in each hand will still be experienced by the same conscious self: Tat tvam asi, even though Prof. Gazzaniga could prove that your awareness was compromised.

Awareness

Sorry to give you a headache, but now I need to replace a chunk of your brain. You had meningitis which totally destroyed the visual cortex on both sides of your brain. You are totally blind. Fortunately, we suppose, you live in an age where the damage can be repaired with electronic implants. After your operation your visual *awareness* is completely restored. You run a flower shop and can make beautifully formed, color coordinated flower arrangements again and drive your van through crowded city streets and make flower arrangement deliveries. Your neurosurgeon asks you if there are any residual problems. You say "Yes, I am still totally blind". I have no conscious experience of anything seen like I did before the disease". "Oh, don't worry about that" he says. "Your visual *awareness* is completely, scientifically, electronically restored to the same visual awareness you had before. You see, electronic circuits don't experience *consciousness*". We can imagine the disease spreading so that you loose your hearing, sensation, taste, smell, emotion, until your experience of reality is that of a long dead corpse or the paperweight on your desk. But electronic neurosurgery has come to the rescue and now your entire brain is now one big silicon crystal. You pass the Turing Imitation test with flying colors and your loved ones are very happy with your complete recovery. If you think there is something screwy about this, so do I. Then why waste time talking rubbish? The above rubbish is exactly what science today tells you is real. As far as science is concerned, awareness is demonstrable and reveals that something is real, but lets not get into the arguments between quantum mechanics Nobel prizewinners Einstein, Bohr, Heisenberg and Schrödinger about what is "real". Dirac stayed out of the fray. Science has nothing to say about consciousness – as far as science is concerned consciousness does not exist. Science is right. It does not. If you put your hand back on the stove burner you will discover that consciousness is real. Science has no way of accounting for how it felt. Schrödinger: *It* (consciousness) *cannot be accounted for in terms of anything else.* Awareness is a theory about what went wrong and that theory is riddled with metaphysics.

Science <u>can</u> study awareness. We can study the cognition mechanisms of the brain with the objective of being able to predict what comes out of the brain if we know what went in. Thus the mechanism of awareness is at least potentially knowable. We may be able to predict that an observable pattern of neural firings is associated with pleasure experienced consciously, such as a cat smelling catnip, but such scientific knowledge does not allow us to experience it ourselves unless there is a way of reproducing the identical cognitive processes in our own brain. For example, we may be able to observe the patterns of neural firings in a spider's brain when she is aware that a fly is caught in her web and to predict how she will behave, but it is scientifically impossible to know what it feels like to be the spider - unless you are the spider in which case it is scientifically impossible to convey to me what it feels like to be you. Unless I am you. The same you that is the cause of the Samkhya paradox if consciousness and awareness are confused.

Consciousness vs Awareness

At least in English, these words often are used interchangeably. One may hear the expression *political consciousness* where political awareness is intended in the sense in which I have used that word. It could mean the conscious experience of emotional fervor inspired by political idealism. It may mean the awareness that the party mathematically does have just enough votes to pass a desired measure. My emotional fervor is not scientifically demonstrable except through empathy, but I can demonstrate the poll results that show my awareness mathematically is correct; a computer program can do it for us.

One cannot rid the quantum mechanics of metaphysics unless one separates the concepts *reality* and *existence*. If one identifies *reality* with *consciousness*, one needs a separate word where I have used *awareness*.

To illustrate the difference between these concepts, one can liken consciousness to the ink with which awareness is printed. The printed material may be profound truth, an approximation to the truth, fantasy, nonsense, pernicious falsehood. However, whether the ink prints truth or falsehood, in all cases the ink is equally real. To one who believes that awareness has a reality independent of consciousness, I can hand a sheet of paper on which awareness of truth is printed. When the recipient complains that the paper is blank, I tell them that I did print the truth on it, but the printer ran out of ink. Awareness without consciousness is void. The print of a printer empty of ink is not real. Without consciousness, awareness has no reality.

Awareness of Consciousness

Most of the time we are not aware that we are consciousness. In the market, we pick up an apple, look it over, toss it back and pick a better one without bruises or bad spots. We see the apple as a physical object in 3D space. We pick up a melon and smell the stem. If it is fruity and sweet we know it is ripe. Here we are using consciousness which does not fit into a physical world model, but know from experience the it is a measure of awareness of ripeness. Only if it is pointed out to us that every thing that is real to us is present in consciousness, to we become aware of being conscious. We cannot assume that a monkey has this ability.

Roger Penrose

Most people do not distinguish consciousness and awareness. In higher brain function they are intimately intertwined. Prof. Gazzaniga did not make this distinction in his famous 1967 split brain paper in Scientific American. A notable exception is the Oxford mathematician Prof. Roger Penrose. You should not read this book without listening to his YouTube presentation *The quantum nature of consciousness.* Penrose makes a sharp distinction between consciousness and awareness.

Future scientific understanding of consciousness will be a bridge between the quantum mechanics and consciousness. It is not the purpose of this book to build that bridge, but to purge physics of the needless metaphysics of Copenhagen and von Neumann's Process I. The title originally was *Physics sans Metaphysics.* In the course of writing it, I came to the realization that this purge had not happened in the better part of the century history of the quantum mechanics, not because physicists were not smart enough, but for a very different reason: They recoil before the word *reality* as does Count Dracula before the cross of Christ, remaining trapped in the only reality they trust: The Physical World (Satan in the Count's case). Yet they have never seen the physical world except as it is clothed in consciousness. For this reason I changed the title, replacing *Physics* with *Reality.*

Here is an illustration of the stark distinction between consciousness and awareness: When we look around, we see a 3D world clothed in consciousness. We also see a 2D subset world when we look at a picture. We are surprised when seeing two such pictures, on for each eye, in a stereoscope: A 3D world springs to life. In a 3D movie, things come out of the screen and float into the theater. On everyone's bucket list should be listening to binaural sound with headphones. It is not equivalent to stereo which is best done without headphones. I first heard binaural in an audio store years ago. The demo announcer told me I was listening to the same flat monaural sound I had heard all my life with headphones, saying in a moment the equipment would switch to binaural. While waiting, a woman behind my right shoulder asked me what I was listening to. I turned to speak to her but no one was there. She laughed over my left shoulder and said "I'm here". I spun around but she was not there either. The brain also constructs the sense of spaciousness from perspective in the two dimensional image on the retina, or phase difference of sound in each ear. This sense of distance is a consciously experienced construct of higher brain function. Humans cannot see what is behind them, but we have a blind-sighted sense of what is there, filling 4π steradians of the two dimensions of directional space. If we magnify or demagnify the image we destroy perspective. We can shrink 4π steradians to a small circle and make it appear as if reflected in a silver ball. We cannot magnify it or it would overlap on itself. If we look at the moon with binoculars, it looks as though we were eight times closer, but the celestial sphere is still only 4π steradians. Our conscious experience is that of a perfect mathematical model in consciousness that is continuous and not limited in awareness by being pixelated or composed of retinal rod and cone data. A star like Mizar in the Big Dipper looks like a point. We cannot imagine what it would feel like in consciousness to see inside that size-less point: to see Mizar with the eyes of a 6 inch telescope and see that it is a double star in a sky that still fills only 4π steradians. The mathematical model our brain constructs is not capable of such a model. What is beneath its power or resolution is imagined as a point that cannot be enlarged to see what is inside it.

The story in awareness, as Prof. Penrose points out, is very different. If we look at a printed word or anywhere near it we can easily read it. If we look a few degrees away we no longer can. If we look a page away we can still be aware that it is a page of print, but can no longer separate the lines. If we look at a fingertip at arms length we can see our finger print, but if we splay out out fingers and look a radian away we cannot even count them. Yet we can see that each finger is in the proper place in the celestial sphere if we wiggle them independently. It is easy for us to know what it feels like to be an insect that sees a wide solid angle with eyes that only have a few thousand pixels. It is what you see a radian from your central vision. In consciousness it is a continuum with no pixelation. In awareness you could not resolve the petals of a flower unless it was very close.

What we consciously experience is cleverly designed to show us what is "really there", without burdening us with the conscious experience of the enormous limitations of visual awareness imposed (probably) by the limited bandwidth of the optic nerve.

Reality and Existence

If I see diamond palaces in the clouds, that conscious experience is **real** whether or not the diamond palaces **exist**. I should really say: "The conscious experience of seeing diamond palaces in the clouds is **real**", which makes no claim about the **existence** of diamond palaces. Thereby I allow the experience to be an hallucination: neural firing patterns not caused by diamonds in the sky, but LSD in the blood. This book excludes the following implicit definition of **reality**: Horses are *real* animals made of atoms, aren't they? But Unicorns are mythical and not *real*. Instead we would say horses exist, etc. This is pure semantics of course. Until God writes the dictionary, people will use words any way they please.

What do we mean when we say: **The physical world exists?** The only thing we can honestly claim to know is that we make observations and we have theories that represent the data we observe. The only thing that exists with any certainty is the theory, but even that is questionable if a mathematician looks askance at it, finding that it is shoddy and lacking in rigor. We ask him to fix it and if he can and it still represents data, confidently we breathe more easily. The physical theory really does *exist* mathematically, even if it is an approximation. *Reality* created in its image an illusion even if it represents conscious experience satisfactorily. If you run into Berkeley, bishop of Cloyne, please tell him this.

The Physical World, programmed into our neurology and underwritten by classical physics, is a very accurate and reliable theory of consciousness. It exists as a mathematical theory, and is quite simple in its geometry. It explains physical objects and explains the conscious experiences you will have when you observe them. For example, you have a 1930's cigarette tin with a painted label, a bit scratched and bent, but it still opens and shuts. You keep a collection of rare coins in it. Before opening the tin, you can predict the conscious experiences I will have examining a certain coin you take out and show to me. The whole game of looking at your rare coin collection plays out in consciousness, but consciousness is not a thing you can keep in your box, because consciousness does not exist. Yet we think of the tin as being ultimately real because it correctly predicts our experience which is never real except in consciousness. The physical world becomes "real" when we create the physical world in the image of the theory. This is equivalent to believing that it is impossible to observe data that can ever contradict the theory. Hume warned: No amount of data can prove a theory true; it takes only one datum to prove it false. The physical world exists as a perfect mathematical theory, but it is wrong in representing experience: It came to an end in the failure of the Michelson/Morley experiment and the ultra-violet catastrophe. Only after the theories of relativity and the quantum mechanics were at hand did we understand why classical theory worked so well: It was, like the flat earth, a very good approximation at low velocity and high mass which is true of the vast realm of common experience. Most of the people in the world believe that the Physical World is ultimate reality. Before the twentieth century, this view was defensible.

Creating reality in the image of theory

What do we mean when we say "The earth is flat". We should mean is: "We have been using a flat earth theory since father was a boy and haven't run into any problems so far. Our contractors and architects always use it and our buildings do not fall down". What we should not mean is: "Our theory works because it is God's truth. That's how God created the world, and I know I am right because I believe in God". That is the self-righteous reasoning of Satan and a violation of the second commandment telling us: *You shall not make for yourself an image in the form of anything in heaven above or on the earth beneath or in the waters below.* The worship of belief is not equivalent to faith in God. Metaphysics has a long history of disobedience to this commandment. The flat earth theory is wrong. Aristotle's dynamics is wrong. Newtonian mechanics is wrong. Schrodinger and Heisenberg's quantum mechanics is wrong. Dirac's relativistic quantum mechanics is not wrong. Yet. In case it turns out to be wrong too, how can we ever comprehend reality? The first thing not to do is to go on making the same mistake believing our theories reveal to us a vision of Absolute Reality. Theories represent experience. They do not reveal God.

The thrust of this epistle is to point out that reality is demonstrated to you every second of your waking or dreaming life: consciousness. It is not something to believe in: Its reality does not dance attendance on your beliefs. We cannot create consciousness in the image of anything: The reality of consciousness is a miracle. If we accept consciousness as reality, Everett shows us that consciousness is infinitely vaster than demonstrable awareness which, apart from the fragmentation of awareness at the level of brain function across a plurality of beings, is limited in scope at the most fundamental level of physics. An even greater limitation appears when we extend awareness to astronomical space and time.

Elementary particles

We have to reserve the word *existence* to describe the elementary particles: leptons, quarks and bosons, without falling into the trap of seeing them as elements of the world of classical physics. Falling into this trap is to believe that they are validly described by the intuitive view that they are elements of the world created in the image of the $c = \infty$; $h = 0$ world. It is this interpretation of the word *existence* that forces upon us the absurdities of the Copenhagen Interpretation: Random causality and the need to collapse inconvenient eigenstates. The analysis of events observed in the Large Hadron Collider at CERN is done using the quantum mechanics and not done using classical physics in the image of which the "physical world" is created. The classical realm cannot account for the quantum mechanics: Einstein's attempt to do so was demonstrably wrong. However, many component parts of the CERN accelerator were designed using classical physics, a labor-saving approximation that does not degrade performance.

The word *existence* is also used mathematics which should not cause a semantic collision in this book.

An elementary particle is not an independently existing, unanalyzable entity. It is, in essence, a set of relationships that reach outward to other things - H.P. Stapp   The Standard Model is a set of such mathematical relationships between particles like electrons. We get into trouble quickly if we define an electron to be a point in space time unless we are justified in falling back to a physical world model such as a β particle in a cloud chamber where such an approximation works. We can use Newtonian and Maxwellian physics to model the paths followed by electrons in an electron microscope: it is a justifiable approximation to quantum mechanics in this case: When the accelerating voltage is 10,000 volts, the electron momentum is so high that the spatial wavelength is so small the electron behaves like a physical world particle, yet is it low enough to be non-relativistic which means that we do not need the Special Theory of relativity to understand how it moves. If we gave Newton an electron microscope, he would not be able to see diffraction effects and would conclude that electrons were not be waves but had to be particles, obeying his laws what's more. He made this mistake with light.

A lot about elementary particles still is not understood. We classify them using group theoretic structures that represent experience and have predictive power like the Higgs boson, but there are sixteen constants of the theory that we have to measure. We have no theory that represents them. It is known that slight changes in these constants would make life in the universe impossible, which evokes the thought that they were chosen to make life possible. This is called the anthropic principle. For present purposes we need only state that imagined laws of physics that are not the structure of consciousness do not represent reality. This view needs to be contrasted with the view of a string theorist that 10^{500} string theories, not one of which is in demonstrable correspondence with reality, imply that an omnipotent dolt created so many bungled universes that he got one right just by mistake. This string theorist creates God in the image of himself. If the laws of physics do not represent conscious experience, they are do not represent reality and belong in the wastepaper basket.

Even if we had a "theory of everything" it would not be possible to infer the subjective content of consciousness given physical law. The subjective content of consciousness is thus a miracle. An example of such a miracle is what it feels like to be a cat smelling catnip. The objective description of what every quark, lepton and boson are doing in a cat's brain cannot make you know what it feels like to be a cat.

To say that lightening was once believed to be a miracle but now science has explained it so it no longer is miraculous, misses the point. What happens to you in <u>consciousness</u> when you see lightening flash, if you live to tell the tale, <u>is</u> a miracle regardless of your understanding in <u>awareness</u> of the physics of lightening bolts.

Before we say *the physical world exists* we need to grow up first. Here is how: When we say *the world is flat* we mean that we can use a flat earth theory for many practical purposes without giving up the understanding that it is not flat. When we say *the physical world exists* we can use many theories including relativity to make practical judgments without giving up the understanding that reality is with us in consciousness throughout life; we need no other reality to understand what the quantum mechanics is talking about. We cannot use the physical world to understand quantum mechanics without absurdity. But we can use it to remember the way to Granny's house for which we definitely do not need the quantum mechanics or relativity.

We cannot find the theory of everything by looking inside the human brain. The widespread beliefs that the earth is flat, the human race descended from Adam an Eve, the sun orbits the earth, the lunar landings and the holocaust are hoaxes, attest to the abysmal stupidity that can possess the average human brain lubricated as it is with superstition; pseudoscience like astrology; believing Wi-Fi causes cancer. Before looking down ones nose at this stupidity, one must add to this list Copenhagen's random and superluminal causality. The neural firing patterns of the human brain is the last place to look for truth; perhaps the worst place in the known universe; even worse than the brains of animals. Animals are delivered from this stupidity by not having the capacity of our intelligence. The first thing man did with powered flight was to loop the loop. Any bird can loop the loop, but none has ever been stupid enough to do so. Flies do, but only to stand on the ceiling. At the bottom of the abyss of human stupidity is a guy, described in the Bible and Koran as Satan, who proclaims his beliefs to be those of God. He appears in several places on every block, on some university lecture podiums, in many church pulpits and and every legislature it seems.

Reality and Science

It is widely believed by scientists that talking about reality is not the job of science; that science is about how things work, not why they work; that talking about reality is the job of philosophers. Philosophical contemplation has never revealed truth that has been successfully determined by experiment.

Philosophy may teach us how to think, but not what to think at least when natural law is sought. Before the last century, this had been true for two millennia, notwithstanding attempts by philosophers to deduce natural law by contemplation. The past one hundred years show that ignoring reality it is not valid science; not in the sense that philosophy can replace experiment, but in the sense that unconscious instinct about what is real makes inroads into the intellectual understanding of how things work. It not only corrupts physics but also the attempt to understand consciousness, by assuming consciousness is a phenomenon: This is a mistake Schrödinger did not make.

The two thousand year road ended, not in the theories of relativity, but in the quantum mechanics when it ran into a century long metaphysical bog where understanding reality is still stuck. For the first fifty years of that century, attempts were made to reground quantum mechanics in "physical reality", like that of EPR. In the next half century it was recognized that doing so was impossible. Prof. Leonard Susskind (and perhaps others not known to me) lucidly shows that Boolean logic applied to the states of a classical system are not valid when applied to the states of a quantum system; that our neural makeup locks us to a visualization of classical reality. Nevertheless he, like most physicists, accepts as physics two elements of the Copenhagen Interpretation pertaining to a system having non-zero amplitude to be in more than one eigenstate of an observation:

- quantum physics is inherently unpredictable in the sense that the observed state being real is randomly chosen, from the list of potentially real observable states, to be the only one that is real.
- all eigenstates of the observation are real until the randomly chosen one is observed which collapses the reality all the others

instantaneously everywhere, unless, of course, they are needed to cause interference.

The first implies breakdown of causality; the second implies instantaneous causality across space, to which Einstein vehemently objected. **We will show that both of these elements are pure metaphysics by showing that neither is necessary to understand the mathematics of quantum mechanics, nor to understand the results of experiments.** They are artifacts of the inability to stop believing in the reality of the "physical word" even while paying intellectual lip service to its demise. **Abandoning it as the homeland of reality evidently cannot be done if one sees no alternative:** For this reason it is not good enough to say that it is <u>not</u> the business of science to contemplate reality, because people anyway, even learned savants, will posit their beliefs in some naive conception or another of what they instinctively think is real. Once there, they will proclaim their view to be rationally grounded objectivity, untainted by philosophical gobbledygook, relegating discussion of reality and consciousness to mysticism, not physics. Wigner, early history of the quantum mechanics, did give serious consideration to the role of consciousness, but fell into the trap of believing the human brain, seen as the source of consciousness, was a crucial element in quantum interpretation. Feynman correctly put paid to this view. Over the objection of some of its founders, notably Einstein and Schrödinger, most accepted the Copenhagen interpretation: At date, nothing has changed in this respect in the first ninety year history of the quantum mechanics. Einstein fell into the "objective reality" trap, seeking recluse in ideas that eventually were discredited experimentally. Schrodinger came much closer, but died too soon.

Three true stories

1. A Hungarian mathematician wrote a letter to his friend the great mathematician Carl Gauss at Göttingen, pointing out that his son Janos Bolyai developed logically consistent geometry can be built on an axiom that parallel lines cross. Gauss wrote back to the effect that he had thought along those lines himself, but decided not to publish them to an unappreciative world who may consider them madness. Gauss did not have the courage to publish what he knew to be true. Bolyai did and got credit for the ideas.

2. My mother was a doctor who worked in a number of mental hospitals. In one of them her supervisor told her the first test of patient's sanity was to ask them the date. If they didn't know it, they were mad. If not knowing the date is insanity, how much more insane are you if you

think parallel lines cross. Gauss had reason to fear.

3. My grandfather was a research scientist with expertise in classical mechanics. He lived to see the quantum mechanics, but did not believe it. To him the physical world was God. If I were to tell him I don't believe that there is any such place as the physical world, he would have believed me to be demented: Like telling the Pope that God is a delusion.

No physicist worth their salt today will claim that classical physics is anything but an approximation to the quantum mechanics and hence reality seen as the classical world is an illusion. This is an intellectual view of what reality is not. It is not a view of what it is. If you are not willing to look at what it is, your gut-level neurology will tell you that you still are in the "physical world" somehow. If you accept this, however unconsciously, you have no choice but to accept the Copenhagen Interpretation.

Here is a simple experiment to determine what reality is:

Apparatus: A particular configuration of atoms. In this case it is you.

Procedure:

• Open your eyes and look around. What you see is conscious reality. What you think you are looking at is your theory of what is real.

• Listen. The sounds you hear are conscious reality. What you think you are listening to is your theory of what is real.

• Reach out and touch something. What you feel is conscious reality. What you think you are feeling is your theory of what is real.

I will not ask you to take the fourth test, but ask you to remember what it felt like the last time you put your hand on a hot stove burner. That is reality and do not forget you are a configuration of atoms, which are not, as the quantum mechanics has taught us, elements of Einstein's "physical reality".

In order to posit consciousness as the reality that the quantum mechanics describes, we <u>have to talk about what it is **not**</u> before talking about what it is. Not doing so is an easy way to fall into philosophical profundities that half of us proclaim exemplars of wisdom and the other half proclaim to be incomprehensible rubbish. I am not putting such people down: This is exactly what happened in mathematics. Poincare ended up on both sides of his own fence. Or falling into metaphysics and gobbledygook: "It is ectoplasm". "A mysterious phenomenon" "It is a scientifically observable meta-substance with transcendent properties", etc.

We have to talk about consciousness by presenting what, for lack of a better word, I will call its axioms that we can use to understand the outcome of quantum mechanical experiments.

The axioms of consciousness

The list may be redundant.

1. *Consciousness is absolutely fundamental* - Schrödinger. Equivalently, consciousness is reality
2. *Consciousness cannot be accounted for in physical terms* - Schrödinger
3. *Consciousness cannot be accounted for in terms of anything else* - Schrödinger
4. The above three axioms are equivalent to the assertion that the subjective content of consciousness is a miracle
5. The binding of conscious states is awareness
6. Consciousness does not exist. This is equivalent to the statement that the physical world is not real, if we understand existence to be the reality of the physical world. Hence the axiom of von Neumann that "consciousness is in the physical world" is false.
7. Consciousness is experienced but cannot be observed. For this reason consciousness is not scientifically observable.
8. There is no scientific proof of the reality of consciousness short of partial brain transplants which call into question who you are in awareness.
9. You are the proof of the reality of consciousness: *Tat tvam asi*. There is no scientific proof that you are conscious. Knowing that you are is proof given to you by God that you cannot give to others.
10. There is no reason to believe baboons do not experiences consciousness. A baboon may not be aware of being conscious but nevertheless is self-aware.
11. Conscious self hood is not denumerable or demonstrable unless you are the experiment. There are not two or more conscious selves. *Tat tvam asi*. Whether there is Zero or One conscious self is a question to be put to one's Zen master, who may answer "What difference? Consciousness is Absolute Nothingness, of which there can be only One".

12. Solipsism: An incontrovertible philosophy that no one believes. The solipsist believes he is the only being in the universe who is conscious. All the rest of you are Turing Imitation Game automatons who act out the role of being conscious, but who experience nothing. No mother who has experienced pain and hears her child crying out in pain can be a solipsist. Yet we cannot throw solipsism into the crank case of crackpot ideas. The reason is that consciousness is not scientifically demonstrable, while none can deny its reality.

13. States of consciousness that can possibly be experienced by the conscious self are real independently of space and time. This is to be understood in the sense that a state of consciousness experienced in the context of some place and time does not cease to be real at others. This negates von Neumann's claim that consciousness is "in" the physical world, hence passing out of reality with the passage of time. If you can remember some significant conscious experience that happened to you as a child, that conscious experience is just as real as the consciousness you are experiencing now. The arrow of time is in awareness, not consciousness.

Consciousness, which forces us to known that it is real, does not exist in the sense that physical objects are said to exist. But consciousness is always associated with events in the realm of the elementary particles which have properties that are mathematically exact. Asserting that Absolute Nothingness is reality even if it is manifest as real consciousness, may appear to say nothing useful, but it does say this: Reality is not what we think of as being the physical world, the homeland of long ton weights, surveyors chains, pendulum clocks and steam locomotives. The paradoxes of the quantum mechanics disappear when we get rid of the physical world. With it gone, the only thing we are left with is consciousness. It may not exist, but it is absolutely real. We exist as configurations of elementary particles which are mathematically exact. If numbers do not depend on physical objects to get their properties, where do they come from? Their properties are apodictic. The physical world is not real, but numbers still have their infinitude of properties which do not owe their mathematical existence to beads, pebbles or bananas. The physical world exists as an approximate theory of experience. Where the approximation breaks down, it is catastrophically wrong. We are the apodictic consequence of Absolute Nothingness. It may appear to the religious reader that this reasoning denies the reality of God. It would if consciousness were not real. Absolute Nothingness is consciousness. The reality of consciousness thus is a miracle.

The doctrine of psycho-physical parallelism is not included in the list of axioms above because, historically, it implies the coincident reality of the physical world and consciousness. We reject the notion that the physical word is real, so we do not need a doctrine that consciousness is parallel to anything. Our neurology, using a tiny fraction of the leptons, quarks and bosons that dominate our experience, constructs a $c=\infty$, $h=0$ model that represents our experience of all the leptons, quarks and bosons that do dominate our experience. The predictions of the model are very accurate for everyday purposes. Its clothing of appropriately subjective conscious states: sight, sound, taste, smell and touch, each multidimensional (color, pitch, hot, cold, sweet, sour, etc) is chosen in ways incomprehensible by present day science or philosophy. These conscious states bind in awareness and are the reality of the model our neurology constructs. Before the quantum mechanics, we could separate the conscious states of the model from the imagined external world machine "the physical world" causing these states, and include our own bodies and brains in that world.

This suggests a redefinition of the doctrine of psycho-physical parallelism to be: *The structure of consciousness, namely awareness, is the quantum mechanics, which cannot generally be framed in physical world models, except when approximation is justified. Where such approximation is justified, it is to be considered a model, and not reality.*

The properties of awareness

1. At the elementary particle foundation, the structure of awareness is Hilbert space. For brain function and many scientific and engineering purposes taking Planck's constant as zero, and the velocity of light infinite, the "physical world" approximation is accurate.

2. Awareness is the binding of consciousness.

3. Awareness is scientifically demonstrable

4. With no proof but the evidence of experience, we proclaim that states of awareness can be, but not necessarily are remembered as elements of the classical realm. Although we could, we do not need to use the time evolving quantum mechanics to understand how. The existence of states of awareness is thus demonstrable, where the word *existence* has the same meaning as the existence

of elementary particles

5. In events that are described by the quantum mechanics, an observed system is consciously experienced to be in all eigenstates of the observation for which the amplitude to be in each state is not zero. This consciousness need not, but may involve brain function. An alarm clock may be conscious without being aware that it is, or that it even exists.

6. Each eigenstate of the observed system causes a state of awareness to be in that eigenstate. Some state(s) of consciousness correspond to each state of awareness experienced for example by a piece of wire, a physicist, a heard of elephant seals. These states of consciousness do not bind in awareness across orthogonal eigenstates.

7. States of awareness caused by different eigenstates of observation are not consistent, which is manifest by their not binding in consciousness. States of awareness caused by the same eigenstate are consistent. This consistency will be observable if such awareness arises in separated places subsequently linked by information transfer. This explains *entanglement.*

8. The density of states of awareness in consciousness is the Born "probability" or the product of the state amplitude by its complex conjugate.

9. The mathematical theory of awareness was given by Hugh Everett in his theory of observation. See chapter IV of his original paper and chapter 5 of his thesis reprinted in *The Many-Worlds Interpretation of Quantum Mechanics* edited by DeWit and Graham. on pages 63ff and 144ff respectively. It assumes knowledge of the math of wave mechanics and is presented in von Neumann's notation, not Dirac's.

10. Self-hood in awareness is demonstrable and denumerable. The set of states of awareness that bind your conscious states define your self-hood in awareness, which is demonstrable: e.g. You can prove you can speak French, drive a car and can climb Mt Everest. I cannot prove to you that I am anything but an automaton that is void of conscious self-hood. Selfhood in awareness is the collection of states of awareness that can bind in consciousness. Thus the boundary of selfhood is where binding ends. If you see me reaching for the sugar bowl just beyond my

grasp, and you hand it to me, this is the experience of the conscious self, *tat tvam asi*, where the binding in awareness is eternal to our brains. When we go our separate ways, this binding ends and we become separate selves in awareness.

The Brain

Assumptions about the role of brain function in the above lists have been avoided because they are inessential to the thesis of this book.

This book argues that the brain is <u>not</u> the source of consciousness. It is the source of awareness. The human brain is capable of being aware of consciousness. One may doubt this of a baboon and confident that a bee is not so aware. Intense conscious states may be real when a door buzzer buzzes, arising from its electromagnetics, but the buzzer itself is certainly aware of absolutely nothing.

<u>The brain is a sophisticated mechanism of awareness, which, at least in principle, is scientifically demonstrable. Awareness is made real by consciousness **not** because the brain creates consciousness, but because **there is no reality except consciousness.**</u>

The important difference between consciousness associated with brain function and consciousness associated with transistors, electric motors, chemical reactions and so on, is that the brain can be aware that it is conscious, although, most of the time, people and probably all of the time animals and insects with brain awareness, are not aware of consciousness. Elsewhere I argued that the subjective content of consciousness played significant roles in the evolution of the brain of the honeybee, not speak of every other brain. Its creator made sure of this, even if the recipient of the gift was too stupid to understand what they were being given. Please see the sub-chapter *The Honey bee* earlier. The first historical account of color blindness in 1777 suggests that, earlier, people were not aware of the consciousness of color vision even while demonstrating awareness of color and certainly were consciously experiencing it. It took from the time of the Big Bang to 1777 for people to give their first dim thoughts to the value of the priceless gift they had been given without the benefit of their advice.

The following comments about the relationship of brain function, brain anatomy, discussion and awareness are not based on scholarly research on my part because there is none. I am flying by the seat of my pants on wings of elementary understanding of brain function and anatomy. I would be happy to stand corrected by anyone with a better understanding.

The gray matter part of the brain surface composed of neurons and synapses are the mechanisms of elementary awareness which are bound to one another by the white matter axons in the brain interior. There may be a hierarchy of such binding of considerable depth. An example of elementary, in the above context, would be in the case I gave earlier of the continuous oscillation at 13.7kz of the tinnitus part of my auditory brain, which is not aware of itself, but activity of which is manifest in consciousness. The other parts of my brain that are writing this sentence, my ego = I, do not bind in awareness with this non-stop oscillation so I, = self-aware ego, does not hear it most of the time, but does so now that I am thinking about it. What changed is binding in awareness and hence awareness of consciousness.

The cerebrum is responsible for thought and sensation and with its extensions like the optic nerves and eye retinas the modeling of space and time which tacitly makes the velocity of light infinite, the non-relativistic approximation, which separates space and time which are modeled separately, not 4D Minkowskian model which would make the Special Theory of Relativity intuitively obvious. In higher brains, space is a 3D flat Euclidean model. It has been reasonably speculated that the brains of sighted insects model space as 2D Riemannian; in blind earthworms it is 1D. Flying towards a flower for a bee means flying long enough to make it big enough to stand on. Wasps interrupted when going through seven purposeful steps to stash food, cannot resume where they left off, but have to go back to step one. Hey, I am no better: Asked for the last four digits of my social security number, I have to start at the beginning.

The cerebrum farms out to the cerebellum dynamic modeling of muscle controlled body function. It makes the tacit Planck's constant = zero approximation: classical mechanics. It has no internal feedback to the cerebrum: Conscious experience of its activity does not bind in awareness with ego of the cerebral cortex. It needs none, because feedback will be external. If it makes a mistake, the pianist will hear it and take corrective action next time.

One can only afford to stop talking about reality in science when one <u>knows</u> that reality is consciousness. The last sentence is not a pontification about reality. If you are not aware you are conscious, I did not address the sentence to you. Sorry, IBM.

Observation: Classical vs Quantum Mechanics

Any serious consideration of a physical theory must take into account the distinction between the objective reality, which is independent of the theory, and the physical concepts with which the theory operates. These concepts are intended to correspond with the objective reality, and by means of these concepts we picture this reality to ourselves.

The above is the first paragraph of Einstein's, Podolsky's and Rosen's 1935 paper (PHYSICAL REVIEW, VOLUME 47) proclaiming the incompleteness of the wave-function as a descriptor of reality. These words are in eminent correspondence with the classical view of the objective existential world being that objective reality seen as the ultimate reality that the laws of physics are describing..

Contrast the above with these words of Erwin Schrödinger:

Consciousness cannot be accounted for in physical terms. For consciousness is absolutely fundamental. It cannot be accounted for in terms of anything else.

These words imply that consciousness, being absolutely fundamental, is the reality underlying experience. For reasons here given earlier, consciousness does not exist and cannot be the object of scientific experiment and hence cannot be the **objective reality**. One could call it the **subjective reality** and say that is what physics is talking about. This I will not do in the interests of avoiding opening a semantic can of worms that will clarify nothing. Instead, I will just call consciousness: **reality**. In either case the words *...and by means of these concepts we picture this reality to ourselves* we fall into a fatal trap: We cannot picture reality. We experience it in consciousness which we cannot observe. What we can observe is manifest in awareness. The credit goes singly to Heisenberg for seeking a formalism that represented data, without trying to construct a reality model. Einstein never got it. Even Schrödinger, who showed that Heisenberg/Born vectors were Hilbert space eigenvectors, was slow to come to the realization that his wave functions did not predict pointer readings to physical measurables. Born saw the truth more clearly, but his probabilistic interpretation swept in metaphysical interpretation that persists to this day which protects physical reality preconception. We can now see the eigenvectors as pointers to conscious states representing experience of awareness that is represented by a formalism shown by Everett to be directly derivable from the equation $A\psi = d\psi/dt$ where A is a linear operator. This equation was known by von Neumann but delivered wrapped in metaphysical machinery that served no purpose but to protect delusion.

Classical observation involves making readings of observed variables which generally have nothing to do with eigenstates and eigenvalues. For example your nurse measures your temperature with a digital thermometer and reads a number from a numerical display or an engineer reads the pressure of steam in a steam locomotive boiler by looking at a pointer reading. He could just as well read the water temperature in the boiler by reading a different pointer. Classical theory relates these two pointer readings.

In some cases it may be meaningful to describe classical mechanical motion as eigenstates as we have done in Appendix F in describing the resonance of a string in a musical instruments in terms of its harmonics, otherwise called eigenfunctions. A much more complex problem would be to ensure that engine vibration does not induce dangerous resonance in the eigenstates of the frame of a vehicle or aircraft.

In any classical case the theory describes the object system and we assume that relevant measurements, eg pointer readings, can be made of the significant parameters of the observed system in ways that have no significant effect on its performance.

The quantum observation is very different. Here the system state is described in terms of the eigenstates of the observing operator which is a component of the observer. To each such eigenstate corresponds an eigenvalue which is an observed real number. The observation constitutes an interaction with the observed system which leaves the **observer** in an eigenstate of the observation. The observer may be as simple as a single photon which is left, say, in the eigenstate of having been reflected by a mirror and also in and eigenstate of having been transmitted. Both are real, but are mutually exclusive in awareness. Or the observer may be a human brain made aware of these two eigenstates. Again, both are real, but are mutually exclusive in awareness.

The essential distinction between eigenstates in classical and quantum observation is:

- Classical: We may observe a piano string as being in a linear superposition of its eigenstates and are we aware of all of them.
- Quantum:: We are only aware of **one** of the eigenstates of the observed system wave-function when it is expanded as a linear superposition of the eigenstates of the observing operator. If our correct theory predicts only one eigenstate, it is certain that it will be the one we experience. But if it predicts non-zero amplitudes to be in several eigenstates, we will be aware of only one of them and if we quickly repeat the same measurement

before things have time to change, we will be aware of the same eigenstate as the last time. In some cases, like an atom being in its ground state, it will stay that way indefinitely. It may be necessary to destroy that state to prove that the atom was in it as theory predicted. You can prove that a firecracker has good gunpowder in it by lighting it. If it goes bang, it did!

All of the mental contortions the great savants of the quantum mechanics went through and are still going through arise from attempts to understand the previous paragraph. We will show that no matter how vehemently you detest injecting philosophical blather about "reality" into science, how you interpret the above paragraph will be forced upon you, not by what your intellect tells you, but by what your gut tells you "reality" is. We are being asked to give up a belief that it programmed into the central nervous system of all living things that have one. It forces its believer to collapse all eigenstates the instant they know they are aware of one of them, even though it it could have been any one. Even God cannot know which one will be selected to be the only reality, or His knowing would prematurely precipitate the collapse while interference was still possible, while recognizing that interference demands the simultaneous reality of all the interfering eigenstates. Over most of the last century, thousands of smart, brilliant people of physics would rather believe this bunkum that get into an argument with their gut level instincts.

Although basic quantum math is not at all complicated, simpler than the math of classical physics, it is very abstract: Trying to relate it to experience, one can easily loose ones footing. In what follows, I avoid abstraction as much as possible but illustrate the problem with a concrete example of a photon being in a linear superposition of two eigenstates. The first experiment to cast light on this was done by Thomas Young around 1800, using his interferometer with zillions of photons, thought of by Young as a wave, not one photon. What happens with just one photon was not understood until the 20th century. In our example, the photon, say from a laser nowadays, is split into two eigenstates: Not the way Young did it, but the way Mandel did with his interferometer, discussed later in this book, which uses a beam-splitter BS. It could be a half-silvered mirror that lets some light though and reflects the rest.

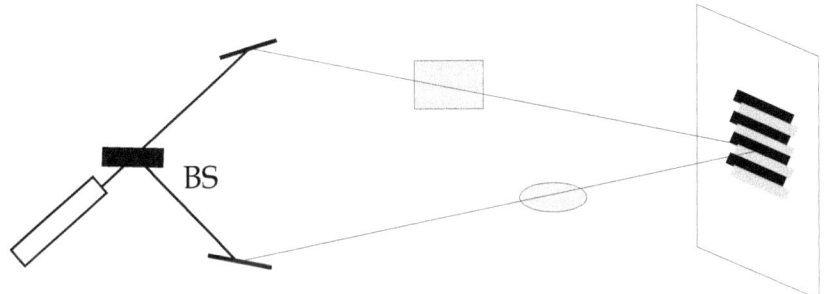

Young's Interferometer with a beam splitter

After being split, the two beams, somehow, say with mirrors, are made to come together again at a screen on the right. On the way the beams may pass through various various optics. Quantum mechanically it is possible to calculate the amplitude for each beam to get to each pint on screen, given specifications of the apparatus. We don't have to worry here how this is done, which in any case is best done by unmarried grad students who will work long hours late into the night enthusiastically for next to no pay, except free coffee. Toiling for endless hours they will calculate the amplitudes for each beam at every point of arrival at the screen. Either beam can cause the photon to arrive at any given point on the screen. If either beam is blocked, the arrival of rate of photons at the screen will be proportional to the square of the amplitude for the unblocked beam; there will be no interference fringes of the kind Young observed.

So what happens if both beams are unblocked? In this case the quantum rule is firm but simple: Add the amplitudes computed by the grad students for each path. This can be done by trained grade school kids who will work for cookies and Gatorade. Because they are signed numbers, they will cancel in the dark fringes and reinforce in the bright fringes. One beam alone will not make fringes. Photon arrival rate as before is amplitude square, or amplitude times its complex conjugate. Your brighter grade school kid can be taught to do this work, but will demand Belgian chocolate and Gatorade.

Whatever one believes the word *real* means, we can say this with certainty: The arrival of **each** photon at the screen **really** uses both paths through the interferometer. Early in the history of the quantum mechanics this gave rise to much mystery mongering by the greatest minds in scientific history as to *which path* the photon took. The very question *"which path?"* is metaphysics because we have just seen that we had to calculate the amplitudes for both paths to understand interference.

Now the quantum mystery deepens: We only save the laser and beam splitter of the above apparatus, but we put 100% efficient photocells in the transmitted and reflected beam paths, and look at photons one at a time, say by turning down the laser so it only emits a few photons a minute.

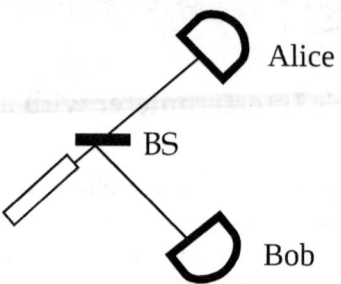

To simplify the description of what we observe, we grab a couple of grad students, say Alice and Bob, and put one at each photocell. As each photon arrives at a photocell, if Alice sees the photon arrive at her photocell, Bob never will. Conversely, if Bob sees the photon arrive at his photocell, Alice never will. Furthermore, whoever whoever observes the photon will get its full h.v quantum of energy, discussed earlier and the other observer gets nothing. Now, instead of the photon taking both paths as it did in the above experiment, it only takes one or the other.

How do we reconcile the results of these two experiments? The attempts to do so over almost a century show that we are at the frontier between physics and metaphysics. The key to resolving it is to stop pretending that it is not the job of science to think about what is real. Let's look at where we are taken by various assumptions about reality

- a. The physical world is ultimately real. We are forced by the interferometer experiment to understand that somehow each photon takes both paths. Why then in the second experiment is it sometimes seen by Alice and sometimes, unpredictably, by Bob, but never by both? Clearly the photon taking both paths just it leaves the beam splitter does not know if it is going to run into Alice and Bob, or go on to rejoin paths at the interference screen. One day, when you go to heaven, you will be able to ask God why sometimes He chooses Bob and sometimes Alice. If we are to believe Feynman, God will tell you He didn't choose anything. Pointing at his dice, He will say "They did. They're boss. I only work here". They keep God very busy, because as soon as the dice choose Alice, God has to stop Bob from seeing the photon by collapsing the eigenstate that put the photon in Bob's path, and which had to be there in case the rest of the apparatus was an interferometer. If Bob and Alice are far apart, God has to work fast: Temporarily suspend the Special Theory of Relativity and reinstate Ancient Egyptian Kinematics to facilitate superluminal causality to get Bob's eigenstate collapsed before it causes any inconsistency in causality.

Why do intelligent people believe this malarkey? Because they cannot stop believing that the physical world is real. They cannot allow both Bob and Alice to experience the photon because the physical world would then be inconsistent. But somehow this inconsistency must be accepted to understand how interferometers work: It was both Bob's and Alice's interferometer paths that really caused the interference.

- b. The hidden parameter approach of EPR: Yes, the physical world **is** real and in it each laser photon is born with a different hidden parameter that predestines whether the mirror will transmit it or reflect it. This theory does not account for interference any way I can see, but I am no Einstein. If he can see a clever way I can't see, it's because Prof. Susskind reminds us: They didn't call him call him Einstein for nuthin'.

If it is an interferometer, we don't have Alice and Bob to tell us, but the beam-splitter must know it is because it now has to let the photon take both paths or we will get no interference. The only theory we need, we already have: The quantum mechanics. An approximate version, like Heisenberg/Schrödinger, is good enough. All we have to do is strip out the metaphysics.

- c. There is another way: Stop believing in the reality of the physical world. "Then, in what reality am I to believe?", one may ask. You don't have to believe in any reality, because you experience it. Let's not have to do the hand-on-hot-stove-burner-experiment again to understand what reality is. The quantum math tells us that the photon enters a linear superposition of two orthogonal eigenstates, both real or we could never see interference. Bob and Alice and their photocells guarantee entry into to states of awareness each caused by one of the orthogonal eigenstates. One is conscious experience of awareness that the photon was transmitted and that it was not reflected. The other is conscious experience of awareness that the photon was reflected and that it was not transmitted These two states of consciousness do not bind in a single state of awareness. They happen to the conscious self, *tat tvam asi,* of Bob and Alice regardless of how awareness is apportioned between them. We are thinking of them as separate people, but bob could be Alice's left brain and alice her right brain and the corpus callosum between them may or may not exist. There are now two equally real histories in consciousness: 1. The awareness of Alice and Bob that the photon was transmitted and not reflected. 2. The awareness of Alice and Bob that the photon was reflected and not transmitted .

Alice and Bob may be very far apart and may never communicate with one another, but if they do, whether at the speed of light or by snail mail, if Bob saw no photon and sends Alice a message, she will receive it in the same state of awareness in which she is aware that she did see it. Equally real is the converse but these two realities never bind in consciousness.

We can see that any state of awareness as a result of being aware that the photon was, say, reflected by the beam splitter, i.e. contingent on its having been reflected, cannot erase the awareness that it did and instate in its place awareness that it did not which would be caused by an orthogonal state. For example we cannot be aware that the photon did something contingent on having been reflected and later watch the same photon do something contingent on it being transmitted by the beam splitter. The only thing we can observe that is caused by both which is a state of awareness caused by addition of both path amplitudes, or interference. Exceptions to this are quantum erasure experiments, to be described later, in which awareness of "which path" the photon took, which destroys interference, can, if caught early enough, be erased, restoring interference. It is impossible to have both interference and awareness of "which path" to the exclusion of the other path because the amplitude for the excluded path would be zero, preventing interference.

If there is no interference, the beam splitter always generates two real journeys in consciousness that do not bind in awareness, because each was caused by an orthogonal eigenstate of the beam splitter. It is not necessary to do so, but easier to understand it the Born weights for each path are equal.   If the experiment is repeated, exactly the same thing will happen. After 10 repetitions there will be 2^{10} remembered pathways in awareness. At the terminal point of each history, the conservation and symmetry laws will be intact. This reality of consciousness bears no resemblance to the presumed reality of the physical world, even though each journey in consciousness seems to happen in the physical world. All this happening to you. Each path is a different journey to your destiny.

Each path is strictly determined, but if you look back along the path that brought you to the terminal, believe metaphysically that is the only reality, then it looks like a random walk through physical reality, engineered by uncertainty. We talk about that next.

The Uncertainty Principle

Uncertainty Principle is unfortunate jargon that got its foot in the door early in the history of the quantum mechanics and has been around ever since. It was blessed by von Neumann who believed quantum physics was inherently random. No one at the time disagreed with with him by arguing that the terminology "*Uncertainty Principle*" was metaphysics.

First, lay metaphysical interpretations aside and look at the solid physics they discovered The physics was Heisenberg's discovery that the operators **P** and **Q**, for the measurement of complementary coordinates, momentum p and position q, do not commute.

$$PQ - QP = i.\hbar.I$$

where I is the unit operator. Nothing is "uncertain" about this equation any more that non-commuting operators in the rest of mathematics is uncertain.

Classical physics demands:

$$pq - qp = 0$$

However the quantum mechanical operators **P** and **Q** do not commute because they have different eigenfunctions. This was discussed earlier. p and q are the complementary pair of coordinates of Bohr akin to the conjugate pairs of Hamilton's pairs of classical mechanics, which requires both p and q to be simultaneously known to solve the equations of motion that determine the history the system, imaged in pointer reading.

In the quantum mechanics if $\psi(x)$ is the wave-function giving the amplitude to find the particle at x, and $\phi(p)$ is the wave-function giving he amplitude to find the particle with momentum p. If these are complementary, these two wave functions are the Fourier transforms of one another. This Fourier transform relationship is strict:

$$\psi(x) = \int \phi(p).e^{+ipx}.dp$$
$$\phi(p) = \int \psi(x).e^{-ipx}.dx$$

If you know one, you can calculate the other <u>exactly</u>. There is <u>no uncertainty</u> here. All of the mathematics above is rigorously true. Either completely describes what is knowable about the system. The proof that this is true in not complicated for those with some math background. See Prof. Leonard Susskind's lecture 9 in the *The Theoretical Minimum*

YouTube series from Stanford University.

For the benefit of the non-mathematical reader here is the look-see math of Fourier transforms:

Here is the time plot of the wave of someone playing a sustained note on a flute, as you would see it in an oscillogram:

This wave goes on forever or however long you can blow the flute before needing more air.

A plot of pitch of sound vs. loudness will look like this:

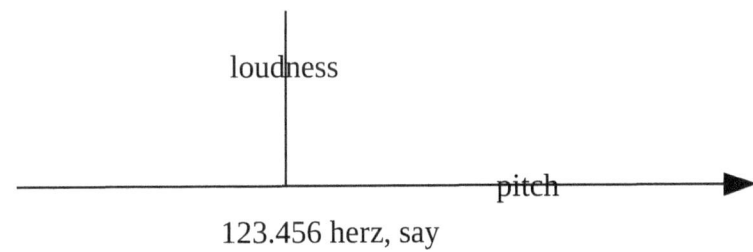

123.456 herz, say

This is called a *spectrum*. Here all the energy is at <u>one frequency only</u>: the pitch note of a flute. A laser generates light like this and it all has one frequency very much higher than a flute note.

<u>The above two functions are Fourier transforms of one another.</u>

If the wave looks like this:

The pitch spectrum will spread out more look like this:

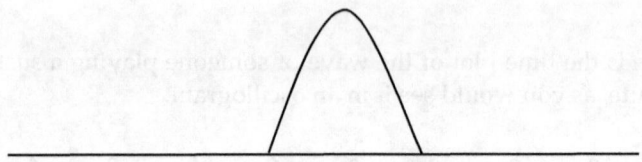

The above two functions are Fourier transforms of one another.

Here the pitch spectrum is a band of frequencies that are all in phase at the center of the band but get out of step and cancel the further you get away from the center.

If we squeeze the wave, say light, into a narrow time window from t_1 to t_2:

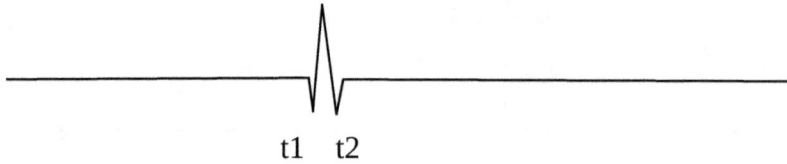

t1 t2

It cannot be represented by one frequency, but takes a wide spread of frequencies, the pitch spectrum will look like this:

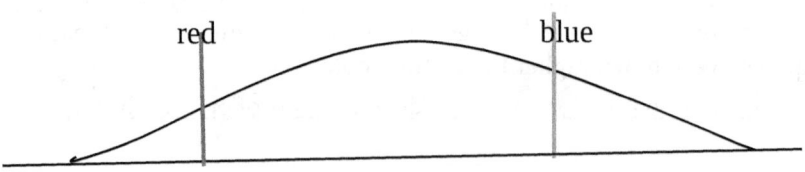

photon frequency or energy

The above two functions are Fourier transforms of one another.

The smaller the time window from t_1 to t_2 the greater the spread in frequency and hence the greater the spread in photon energy if we are talking about light waves.

Quantum mechanically, the above diagrams would depict the strict reciprocity in the spread of the wave functions of complementary observables like position and momentum; time and energy; angle and angular momentum; etc. Even with sound they are complex numbers because both amplitude and phase are important, I have taken artistic license drawing them. The second function in each case as I drew it would be the Born "probability" of the transform where the word in quotation marks reflects metaphysical tradition. I prefer the word "weight".

In classical mechanics, if you want to predict the future history of a system exactly, you have to know every pair of conjugate coordinates exactly at the same instant. Set your clock to zero at that instant and at any other time T the differential equations of motion will tell you the system state: every pair of conjugate coordinates exactly at T.

QM and U are having a chat.

QM: In quantum mechanics a complete description of the quantum system is its state vector Ψ.

U: Ψ as a function of what?

QM: Anything you wish.

U: I want a photon with energy E but I want to know E exactly.

QM: Not a problem. Just use a highly monochromatic laser.

U. I want to control the exact time T the photon is emitted.

QM: In that case you need an energy level with a very high transition probability to the ground state giving you the energy E you want. Press a button that will excite the chosen state any time T you want the photon.

U: You understand of course that I want to to know both T and E of every photon exactly.

QM: You want $\Psi(T)$ and $\Psi(E)$ to be 1 at T and E and zero for all other values of time and energy to be Fourier transforms of one another?

U: Yes, if that's what it takes.

QM. You'll have to talk to Fourier about that. He is going to be annoyed. Be careful: He hangs out with Napoleon who can be dangerous.

The above is mathematical physics. This formula $\Delta T.\Delta E > \hbar/2$ is true of every Fourier transform and simply describes the reciprocity in the ranges of time T and energy E, written ΔT and ΔE for which the wave-functions $\psi(T)$ and $\phi(E)$ are non-zero. These ranges are not at all uncertain. We could just as well talk about the reciprocity between position and momentum, but I used the energy – time case to illustrate a point: In an interferometer, ΔE is related to Δλ the spread in wavelength which has be very tight. If the interferometer is 1 meter long, that is about 2,000,0000 wavelengths and we really want something ten times better, so Δλ hence ΔE must be very narrow; 1:20,000,000. Using natural units in which ℏ = 1, we can calculate ΔT and it must be 10,000,000 natural time units or very long. This simply means that the wavefunction needs a lot of stretch room in time - it is non-zero for a long period of time for each photon. Or, it needs a lot of stretch room in energy if you want the time of transit to be confined to a narrow time window. In the early days of the quantum mechanics, Born and Heisenberg saw photons as classical particles shooting through the interferometer at some exact instant but darn it, if you tried to find out when, you destroyed wave interference. So the time the photon made its journey was "Uncertain" and ΔT.ΔE > 1/2 was "The Uncertainty Principle". If you try to narrow the time window by chopping off its ends, you are stetting ψ(T) to zero there where is was not zero, and you turn the Fourier transform crank and look at the resulting energy hence wavelength transform, it may be spread out over such a wider band of wave lengths that the interferometer will no longer work.

Now, are you ready for some metaphysical math? Here it is, straight from the pages of every quantum physics text of the better part of the last century:

$$\Delta x . \Delta p > \hbar/2$$

Where Δ is called the "uncertainty" in the measured value of x or p. The word "uncertainty" is hogwash. It arose in the early history of the quantum mechanics when everyone with their heads still in the physical world of classical mechanics saw x and p always as numbers having some single certain value and believed the quantum mechanics also did, but was keeping them a secret, keeping you uncertain. They could not understand the simple message: If x is in the range 10 to 90, every value in that range is equally real, but as Dirac showed every value is an eigenstate orthogonal to every other. When they observe x = 56.789, it is because it is real, but so is every other value from 10 to 90. This cannot be true in the physical world or every other value 10-90 also would be experienced. It is true in consciousness because the conscious experience of any value does not bind in awareness with any other: They are caused by orthogonal eigenstates.

This is a statistical formula that correctly describes the results of many measurements, where Δ is the *standard deviation*, a statistical parameter that describes the amount of spread in a collection of measurements. For example, if pistons nominally 10.000 cm diameter coming of a production line were never smaller than 19.998 cm and never larger than 10.001 cm, the *standard deviation* would be small. If a certain adult snake typically 1.5 meters long ranged between 1 meter to 2 meters, the *standard deviation* would be large.

The metaphysics arises in the mind of those who believe that the above formula describes the underlying reality of a single measurement as being inherently random. The adherents of this philosophy include Heisenberg, Dirac, von Neumann, Feynman, but not Einstein. Shortly after Einstein died, Everett showed that the above formula was a statistical spin-off of the strict determinism that is the foundation of quantum mechanics.

Knowing the above does not deliver us from the Uncertainty Principle because even if you do not obstruct the time window but measure the time the photon makes its journey in a repeated series of experiments, the times of transit scatter at seemingly random times in the window. Dirac pointed out that all the times in the window are orthogonal eigenstates. Everett removed the uncertainty by recognizing that all observed times in the window are equally real. He did not take the needed step to recognize that each is equally real in consciousness but these states do not bind in a singe state of awareness. Any avenue we take with the passage of time will seem to happen in the physical word, not because the physical world is real, but because our neurological machinery is programmed to represent conscious experience by an $h=0, c=\infty$ model. It is never experienced in anything in consciousness because nothing else is real; not because our brains are spinning off consciousness.

The humdinger for physics is that we do not observe the above spread with each photon but we see each photon at one frequency and energy, like the gray lines above. Although the quantity observed is one member of a continuum, every member is an eigenstate which is orthogonal to every other eigenstate. All are equally real: A Hilbert space vector points at each. These vectors are all orthogonal to one another. That means if you are of aware of one it is impossible to be aware of any other. Everett showed us that this is what quantum math has been telling us since its get go. If you are not interested in reality, that is the end of this story. Toss this book in the fire and get on with your life If you regret doing so you can always buy another copy later. If you are interested, there are three ways to go:

1. The Physical World is real. But only one can be real at a time. As soon as you experience one for which **p**, or **q**, has an observed value, stamp out the infinitude of others. Now! Everywhere!

Instantly! And tell that squealing brat Albert E. to keep quiet.

2. The Physical World is real. All the **q** in the range **Δq** are real, and there is an uncountable infinity of them in Parallel Physical Universes. Man! That's a lot of mass to have hanging around, but not to worry: God puts each one in its own bag of boson proof anti-gravity wrap.

3. Toss the Physical World in the malarkey crock. Now we have to listen to another brat, John von crying because his Physical Reality toy was taken away. If he wants something real to cry about, hive him a thumbtack to sit on. No, John: pain doesn't come in a physical world box: Reality doesn't exist, but it can hurt, so keep quiet and let mom pull out the thumbtack.

The philosophical pitfall of the metaphysical "uncertainty principle" of the Copenhagen Interpretation is that it presumes that what we be aware of is the only thing that is real, like the red photon in the above diagram. If the blue photon was also real, we would have been aware of it too and these two states of awareness would be consciously experienced as one state bound by both. That did not happen, so, the moment we observe anything all other theoretically real possibilities must be collapsed into non-reality. We cannot say that they never were real, because we need them all or our interferometer would not work. What is real in this case is a continuum of states of consciousness each with an associated amplitude in awareness that is a continuum. When states are discrete we can have a definite amplitude to find the system in some state. When they are continuous, like measurements of variables like energy and time which have non-commuting operators, the amplitude for a variable to have some definite value is infinitesimal. It only makes sense to say that the amplitude to observe the system with some energy at some time is $a.\delta t$ where a is the amplitude density and δt an interval of time. If this is non-zero then consciousness awareness of this state will be real but bind with no other awareness.

In Everett's interpretation, to which we add the assertion that reality is consciousness and not the physical world, continuous states are infinitely many paths in consciousness. Given the number of quantum events that occur in the history of what we otherwise call the "physical universe", this gives God a lot to remember. It may please the religious reader to proclaim that the state vector of the entire universe from the Big Bang to the end of time seen as an isolated system is a model of the mind of God's in which all physically possible paths of awareness are absolutely real. Both Bohm – de Broglie and Everett interpretations say it makes sense to talk of the wave function of the entire universe. The state vector of the entire realm of universes may be very simple. The magic lies in the chains of vector products of eigenstates as causes and conscious awareness of effect that make up all possible paths in consciousness from the Big Bang to the end of time. Because you find yourself on one of these paths, does not mean the others are not just as real.

This is very different view of reality than that of the physical world in which experience unfolds as a random walk incomprehensible to God because we create reality in the image of awareness.

The "Uncertainty Principle" and light

We think of a photon as a particle. Everyone has been calling photons particles for a century. And electrons too: Tiny little specks of charge and mass orbiting protons and nuclei. And when we see one of these tykes, that's exactly what we see – always exactly somewhere at exactly some time, but God prohibits us from predicting the exact spot. Uncertainty, you see! A Fundamental Principle of Nature, you know! Most physicists today believe this.

Quantum math tells us a different story: A photon from a star ten light years away has a wave-front of a hundred of square light years in which the amplitude to be found at each point on that wave-front is non-zero. If your telescope is ten light years in diameter, you can determine the direction of the star very accurately by observing just one photon. But suppose the diameter of your telescope is not ten light years, but ten centimeters. The amplitude for a given photon to be inside that ten cm hole is very, very, very, very, very small. But there are a very, very, very, very, very large number of photons coming from that star, so you wont have to wait long to observe one. Suppose the star is a binary with two equally bright stars as far apart as the earth is from the sun.

With our 10 cm telescope these two stars would look like this:

This image is not a picture of the stars at all. They are tiny dots near the center line of the central blob. The above image is called a Raleigh diffraction pattern and it is caused by our throwing away a hundred square light years of the wave function and looking at only 100 sq cm of it. We can calculate the above image using the quantum mechanics. It is the image of the two stars after we threw away all of the information contained in the photon wave function <u>outside</u> the 10 cm circle. It is the amplitude multiplied by its complex conjugate to see the photon at every point in the above image. It is the Fourier Transform of all of the points in the 10cm telescope aperture which is the photon direction mapped as points in the image plane. Werner says God won't tell us the exact spot our particle went through the 10 cm hole – The Uncertainty Principle forbids us from knowing which exact spot. No! Werner. The photon used <u>all</u> of the 10 cm aperture to reach every pint in the mage plane of the above Raleigh pattern. What we are now absolutely certain of is that the photon did <u>not</u> use any of the rest of the 100 square light year area to get there. If we can see each individual photon arriving, it will appear at random points in the central blob and the diffraction rings surrounding it. Only after a large number of photons have arrived will the above pattern emerge, and it will match our calculations. Heisenberg calls this the Uncertainty Principle, because you cannot predict where a single photon will appear next. Everett rejects this metaphysics. <u>Each photon you see is **exactly** where it was predicted to be</u>: one of the eigenstates of the observing operator which, in this case, is an infinite continuum. In this epistle, each such state of consciousness as the one you experienced for one eigenstate, is equally real for every other, but they do not bind in awareness. Because consciousness does not exist, there is no violation of energy conservation. Energy is not ultimate reality; consciousness is. The conservation of energy, charge, momentum etc., is the property of awareness. By looking at a lot of photons, you are seeing wave function of one, but you still cannot see the two stars because you could not afford to buy a bigger telescope that does not throw away so much of the wave function.

If we understand reality to exist, i.e. the physical world, then the Many Worlds interpretation implies that each arriving photon in the above pattern creates infinitely many physical universes with an infinite increase in total energy. This metaphysics, required by the presumed reality of the physical world, is so absurd that it is easier to believe that each photon arrives at an unpredictable place and the reality of all the other worlds needed to account for the diffraction pattern be collapsed at the instant of observation. This is the underpinning of the Copenhagen Interpretation Werner gave us. Causality of the arrival of the photon is abandoned and pseudo causality invoked to collapse all but the observed eigenstate.

With a 10 meter telescope outside the atmosphere, the two stars would look like this:

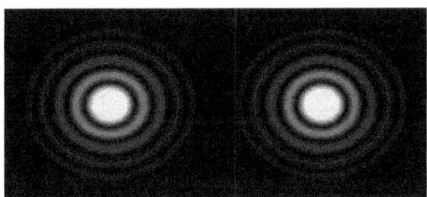

We can now see the two stars clearly separated, but each star image is a Raleigh diffraction pattern and it is caused by our throwing away a hundred square light years of the wave function and looking at only one hundred square meters of it. But we get a clearer picture than we could by only looking at what came through a ten cm hole. Each Raleigh diffraction pattern is the Fourier transform of the part of the wave-function that got through the 10 meter telescope aperture. The rest had non zero amplitude to hit the floor, the surrounding countryside, Mars, to name a few places. There is no uncertainty here.

The following explanation in italics of what is happening is concoction on the part of those who cannot stop believing that the photon is a particle that took some single path to get to the image plane: *We had to give up knowing the exact point in the objective lens or mirror it came from: When we knew to within ten cm, but we could not see the two stars. Now we have given up that certainty by not knowing where in a ten meter hole it came through. Much more uncertain: But now we can tell that there are two stars. This only goes to show how God creates reality using the formula $\Delta p . \Delta q = \hbar$.* You will find this metaphysical uncertainty pseudo-science in many popular and technical books on physics over the past 80 years.

Every single photon we see <u>uses every path</u> through the 10 cm hole when we couldn't resolve the stars, and now every single photon we see <u>uses every path</u> through the 10 meter hole and we can resolve them. This is what the quantum math, not what the meta-physicist, tells us. If you take the Fourier transform of these apertures you will get the Raleigh diffraction patterns shown above. We see Raleigh diffraction patterns instead of stars in both cases because we are throwing away one hundred sq light years of wave-front. What we threw away are the amplitudes that would cancel in our image plane where we are seeing, among other things, diffraction rings. By collecting enough data from multiple photons, we can find where the star is located: At the center of the diffraction rings.

With a 100km telescope aperture (there is no such optical instrument today) one of these stars may look like this:

We can now see a clear image of each star and its sunspots. One of the tiny black dots could be a planet passing in front its star. Because we have no 100km aperture telescope, we have never seen such an image of a star, except for one: The sun, shown in the above photograph. The sun is not ten light years, but only ten light minutes away. The above picture is typical of what we would see of the sun with a ten cm telescope with a solar filter covering the objective.

If our telescope had a ten light year aperture, it would be the mother of all burning glasses and we would be incinerated instantly with the energy of 100,000,000 nuclear bombs by any star we looked at. Pretend we have a good enough solar filter to prevent this fate. We have converted our telescope into a microscope. Now, we could point it at a planet of the star, change filters, and look at microbes on it. But we couldn't see anything much smaller because the wavelength of light is about the size of a microbe.

The only "Uncertainty" in the quantum mechanics is that if you do not look at all that is real, you really will be uncertain about what you certainly ignored. But you can buy back some certainty by looking at a lot of photons instead of just one of them.

One day fairy tale books will tell children about a fantastic place called the Physical World in which the ancients used to believe, where King Arthur lived with Guinevere, legislating Ignorance in the name of

Principle; serving divinely imposed Uncertainty by impaling unwanted eigenstates with his sword Excalibur.

How God created elementary particles

Even if the book existed, understanding how particles got their properties would not be light bedtime reading, but understanding why they are particles is. It was explained by Dirac and Everett.

Suppose you set yourself up with measuring equipment that can measure the position and time of an event both very accurately. We are talking about, say, an electron,. You know its state vector as functions of time and position which are non zero over regions of space and time as its wave nature would imply. Dirac showed that these functions are linear superpositions of Dirac delta functions which contain an area equal to the amplitude but which only have a width which is an instant of time or a point of space for every position and every time and are zero for every other. This mathematical audacity did not go down well with mathematician John von Neumann, but eventually the shouting died down and Dirac was granted mathematical reprieve. The scalar product of two of these delta functions is this zero unless it is the scalar product of delta function with itself. Another way of saying this is that they are orthogonal eigenstates. If we observe the electron in either space or time every eigenstate is a state of awareness, but these states of awareness will not bind in a single state of consciousness. We are left in a state of awareness that the electron in only at a single point at a single time, in other words, it is a particle. Each such state begins a new history in consciousness. That is the basis for particles.

The physical world philosopher can only accept one state of awareness as real and must, somehow, get rid of the others. Admitting that the wave-function covers ranges of space and time, he sees these as "zones of uncertainty" about where the particle "really" is. He is forced to the conclusion that the particle, which always was a particle of course, is chosen by authority baffling to God to be only at the observed place and observed time. Now, knowing where, God's job is to move at infinite speed to collapse all the other eigenstates where the particle must no longer be allowed to be, but had to be to account for interference. Or something. A one word way to describe this philosophy is: Balderdash.

The Foundation of Mathematics

Einstein once said "The most incomprehensible thing about the world is that it is comprehensible". It is amazing that the simple high school algebraic equations of Newton, his gravity formula and some simple equations of Euclidean geometry give a set of equations that describe the motion of the bodies of the solar system to an accuracy of 1 part in 100,000,000. These limitations on accuracy are imposed by observation and not physical law.

This raises the question "On what foundation does mathematics rest". Maybe that's where the foundation of physics rests too.

Summary of the history of mathematical logic

Mathematicians built a logical foundation called Set Theory for the edifice of mathematics based on logical axioms. This work was largely completed by the philosopher Bertrand Russell and Alfred Whitehead and published under the title *Principia Mathematica*. As the work progressed more axioms were proclaimed to be true and added if they gave the desired structure. The goal was to place mathematics on a foundation that was visible and not based on ineffable reasoning that causes mathematicians to disagree with one another. The great mathematician Poincaré even disagreed with himself. It may seem audacious to proclaim axioms to be true in place of proof. The cleverness of what they were doing was not going elsewhere for proof: a backward step. Proof was kept inside the system: Theorems derived from the axioms had to be true if the axioms were not inconsistent with one another.

Proofs of theorems could be built up with a typewriter reserving some symbols for operators and others for variables. The structure was called the Sentential Calculus. A proof based on indisputable axioms and rules of inference was laid out in print for all to see. It was no longer necessary for one mathematician to try to look into the head of another. The axioms were selected to give rise to the desired structure: The structure was visible to every mathematician, allowing them to agree how each theorem was proven with the proof resting on a foundation of consistent logical axioms and rules of inference that were demonstrably not inconsistent with one another,

Alan Turing invented a computer now called the Turing Machine that could be programmed to carry out proof mechanically. He did not intend such a machine to be built, but to show that the process of proof in principle could be mechanized in a way that could not be construed to be based on ineffable reasoning. Discovering the basis of mathematical reasoning could be done by taking a screwdriver to the Turing machine instead of a scalpel to the mathematician's brain. Modern computers were invented later by John von Neumann, but are mathematically equivalent to Turing machines.

Failure could only happen if there were inconsistencies: There was only one way the logical system could fail: If one could prove a theorem true, and starting from the axioms again, perhaps by a more circuitous route of inference, prove it false, all theorems could be proven both true and false and the entire structure would collapse in a heap of nonsense. For a logic to be free of this danger there had to be a proof of its consistency: that no theorem could be proven both true and false.

For a logic strong enough to be a basis for addition in arithmetic, a proof was found in a theorem that was true and demonstrably never could be proven false. This theorem protected the entire structure because if any other theorem could be proven true and false all theorems could be, including this special case, and in this special case was known to be impossible. If the axioms were true by definition, any theorem proved from them hence was true and could not possibly be false.

However, to logically define multiplication in arithmetic, a new axiom had to be added, called the *axiom of choice.* Doing this undermined the proof of consistency. This did not mean that the logic was inconsistent, but a new proof of consistency had to be found. In the absence of such a proof, but assuming it would be forthcoming, Russell and Whitehead forged ahead with *Principia Mathematica.*

Gödel's Proof

Around 1930, about the time the quantum mechanics matured, a young mathematician Kurt Gödel proved that what Russell and Whitehead were trying to do was impossible. Gödel demonstrated a theorem that could be presented in the Sentential Calculus proclaiming of itself that no derivation of its statement in the sentential calculus was possible from the axioms of *Principia Mathematica* or any equivalent system. To believe it was false was unmitigated disaster because then a proof it was true would be possible and *Principia Mathematica* would collapse into a pile of nonsensical contradiction.

At first it may seem not to be a problem: Since the theorem was true, add it to the list of axioms, no one of which can be derived from the others, so why not throw in this one too, since it could not be derived from, or contradicted by, them either. Now it is included in *Principia Mathematica*. There.

Gödel showed that if you did this there would be other unprovable relationships between natural numbers outside the augmented logic. Further addition of axioms would never encompass everything. He proved that no logic sufficient to represent all of arithmetic could contain a proof of its own consistency.

Russell gave up mathematics and took up humanistic philosophy.

Meta-mathematics

It was recognized before Gödel's time by Hilbert that statements about mathematics that are certainly true or certainly false don't necessarily belong to mathematics. The name meta-mathematics was coined to describe that science. Gödel devised a scheme that enabled any sentence or sequence of sentences of the sentential calculus to be encoded as a single integer (whole number). The scheme was a fairly simple replacement code in which a number replaced every cipher. To preserve the sequential order without establishing a convention external to arithmetic, Gödel did this:

raise the 1st prime number   2 to the power of the number for the 1st cipher

raise the 2nd prime number 3 to the power of the number for the 2nd cipher

raise the 3rd prime number 5 to the power of the number for the 3rd cipher

raise the 4th prime number 7 to the power of the number for the 4th cipher

raise the 5th prime number 11 to the power of the number for the 5th cipher

You get the picture.

Then he multiplied the whole mess together and got just one number. This number came to be called the Gödel number of the sentence.

One could easily decode a Gödel number.

• The number of times you could divide it by 2 getting a remainder of 0 was the first cipher replacement code.

• The number of times you could divide what's left by 3 getting a remainder of 0 was the 2nd cipher replacement code.

- The number of times you could divide what's left by 5 getting a remainder of 0 was the 3rd cipher replacement code.
- The number of times you could divide what's left by 7 getting a remainder of 0 was the 4th cipher replacement code.

And so on for prime numbers beyond 7, namely 11, 13, 17, 19, 23.......

Gödel amazing achievement was to prove that every meta-mathematical truth corresponded to an arithmetic property of its Gödel number or a meta-mathematical truth relating two sentences corresponded to an arithmetic relationship between the Gödel numbers that encoded them. A simple example: The meta-mathematical truth "Sentence A is the first part of sentence B". If the Gödel numbers of these sentences were GA and GB then the arithmetic property relating these numbers is GA is a factor of GB. You can see this from the way the Gödel numbers were constructed. So "Sentence A is the first part of sentence B" and " GB/GA leaves a remainder = 0" are two different ways of saying the same thing.

The meta-mathematical truth "Sentence A is the proof of the truth of the theorem represented by sentence B", is a very esoteric arithmetic relationship between their Gödel numbers GA and GB.

Gödel found a sentence in the Sentential Calculus that said:

"The theorem of Gödel number Gx has no proof".

If you decoded Gx, it decoded thus;

"The theorem of Gödel number Gx has no proof".

In other words, it was talking about itself which was a very esoteric arithmetic property of Gx.

This was the proof that resulted in the *Principia Mathematica* falling infinitely far short of its intended goal which was to encompass all of arithmetic and the rest of known mathematics in a demonstrably consistent logic of a finite number of axioms.

What Gödel also did, although it may not have been his intention to do so, was to prove that a foundation of mathematics could not be built out of elements of the classical realm of the Physical World: Ciphers of the Sentential Calculus on printed pages, nor represented in the mechanical forms of Turing machines.

The foundation of physics is in mathematics that cannot rest on a foundation of elements of a Physical World as Bertrand Russell hoped would be the case for mathematics. It is important to understand that Gödel did not show that there is a boundary to logic beyond which lies magic. He rather showed that what appears as magic beyond some boundary may lie within a greater boundary of higher logic, beyond which lies even greater magic, and so on forever.

Gödel's proof was the third blow in a triple whammy that knocked out the "physical world". The other two were the quantum mechanics and the Special Theory of relativity which deposed the absolutivity of mass, length and time.

Russell, Gödel, Grade school

Many are left with memories of primary school arithmetic as being rigid, mechanical and lacking in nuance and subtlety. It was precisely this rigidity that Russell wanted as the foundation of mathematics, free of the curses of ineffability, ambiguity and difference of interpretation. Indeed computers depend on this mechanical rigidity to do accurate and unambiguous calculations. Quantum uncertainty is not an issue here. Although it is a nuisance, machines can be built with arbitrary high redundancy. However, the results of these calculations, freed of quantum uncertainty as they may be, are only approximations to mathematical truth, yielding perfect results only in a limited set of cases. Gödel showed that however great the number of axioms of logic that may be that embodied a machine, there is always an infinite realm of relationships between natural numbers that are beyond the machine's imagination. Within its domain of imagination there is mechanical rigidity. Beyond it, there is magic: Certain truth, yet beyond the machine's imagination. In the war between logical imagination and magic, magic always wins.

The certainty of knowledge we have about elementary particles is mathematical. Then where is the foundation of mathematics? Bertrand Russell learned the hard way that it is not in the $c=\infty$; $h=0$ world of classical physics. It does not lie in the ink marks on paper of the Sentential Calculus or the cogs and gears; the AND and OR gates of a Turing machine. These are physical objects, elements of the $h=0; c=\infty$ world, elements of permanence and immutability that can be used to make a machine without limits on the frontier of its finite imagination, but leaving it with intelligence that cannot comprehend the infinite realm that lies beyond its frontier.

Yet nucleotide bases, four configurations of the atoms each consisting of only a dozen hydrogen, carbon, nitrogen and oxygen atoms can be seen as physical objects, building blocks of the temple of life, which abounds in multifarious forms for billions of years. Quantum "uncertainty" takes its toll: Cancer kills some but it cannot kill all. These molecules are safe foundations for life, but its is factious metaphysics to believe that they get this safety because they in turn rest on words which belong on its epitaph:

Here lies the Absolute Rock of Ages

The Physical World.

The Special Theory of relativity and the quantum mechanics shook its foundation to rubble.

Everett's simple interpretation shows us the illusion of the world of physical objects. The number of paths from the Big Bang to the end of time is infinite. Yet every path is absolutely certain and real. The role of the elemental particles in constructing each path is that, along all paths, the conservation and symmetry laws of physics are valid. If one believes that this can only happen in The Physical Universe, one is dumped into one of two metaphysical bins:

1. The random non-causal metaphysics of Uncertainty and Collapse of the Copenhagen Interpretation.

2. The Many Worlds view of DeWitt and Graham with each event spawning infinitudes of new Physical Worlds.

Dancing on the point of these metaphysical needles can be avoided by not looking for some reality to believe in other than consciousness, which is real no matter what one believes. Because consciousness does not exist, the foundation of consciousness is Absolute Nothingness which requires no foundation. We do need to build a bridge between Absolute Nothingness and apodictic structure. It was shown to us by Gödel.

Elementary particles are mathematically exact

One of the most remarkable events of the 20th century if not of all history, was the discovery that elementary particles have properties that are mathematically exact. Natural numbers do not have a physical existence. The point of contact between elementary particles and mathematics is that all particles of given type, say the electron, are identical in properties with mathematical exactitude. At first, this may not seem to be a big deal. You can count your fingers. That's mathematically exact. The mathematics of physics involves measurements in terms of physical units like the kilogram, inch, second, etc. No two kilogram weights, no two meter sticks, no two clocks that tick once a second are mathematically exactly the same. But two electrons have exactly the same electric charge which is a physically measurable parameter subject to measurement error. Elementary particles of the same kind are also indistinguishable, which became apparent early in the history of the quantum mechanics when it was discovered that they do not obey the statistical laws of physical objects. We do not understand the relationship between arithmetic and the properties of the elemental particles of physics. If we did, we would have The Theory of Everything.

The exactitude described above has its foundation in mathematical truth, and is not bestowed on elementary particles by being part of the "physical world". The physical world inherits its intuitive "rock of ages" permanence and immutability from the mathematical certainty of particle properties. However, the physical world is a flawed model of reality. String theory appears to continue the attempt to create reality in the image of theoretical models. A better place to look may be number theory which does not try to create models of mathematical truth, like those of the sentential calculus. Being a mathematical dunce, I have no idea how to go about doing this.

Bertrand Russell and Alfred North Whitehead and other mathematicians tried to put mathematics on a logical foundation, but Gödel proved that relationships between natural numbers image statements about the structure of any logic based on any number of axioms. These properties of natural numbers are a boundless intelligence that understands logic. The converse is not true. The number $2^{57,885,161} - 1$ is known to be prime. The number of particles in the universe is less than 2^{300}. This prime was not found by playing with arrangements of $2^{60000000}$ beads.

Relativity

We learn at an early age that the world outside our heads is much bigger that the one inside. We are fallible. The external world is not. We cannot remember where we left our keys. The outside world never forgets. It is the rock of ages, immutable in space and time, proclaiming its ponderable eternal immutability.

Until 1905.

Then we are told that if calibrated meter rods A and B are in relative motion:

A measures B and it is shorter than A

B measures A and it is shorter than B

If two calibrated clocks are in relative motion:

A measures B and it runs slower than A

B measures A and it runs slower than B

If two calibrated kilogram weights A and B are in relative motion:

A measures B and it is heavier than A

B measures A and it is heavier than B

Either these statements are hogwash or the "physical world" is not real. Before 1905, the above truths could only be real in Alice's Wonderland. After 1905 they were the stark truth of Physical Reality.

The most spectacular effect of the Lorenz transformations is the accumulated disagreement between a clock moving relative to a system of synchronized clocks illustrated by the twin paradox of the Special Theory. It is described in Appendix G. The Special Theory did much greater damage to the metaphysics of Physical Reality than is commonly acknowledged. The General theory did more.

Emanuel Kant proclaimed Euclidean geometry, to be the ultimate correspondence between the mind of man and the mind of God. Kant had no reason to believe that it was anything but perfect. The General Theory showed it to be an approximation. Kant deserves no criticism for being wrong: It was reasonable to assume in his day that the laws of physics were exact. Today we believe that elementary particles have mathematically exact properties, but where Kant was wrong, we may be right.

Quantum Mathematics Interprets Itself

The key to understanding the quantum mechanics in simple terms is to ask how our awareness is caused to change by observation. Thirty years after the quantum mechanics made her debut, Hugh Everett showed us how in his PhD thesis, condensed from an earlier paper. Everett did not treat the system being examined as an object we can observe from our vantage point in the classical realm where pointer readings are definite and we are observers outside the system we are looking at. This is what the Copenhagen Interpretation did by effectively proclaiming the ultimate reality of the classical world. Everett described <u>you</u> with the wave-function U. The system being observed in this case is described by the wave-function M. In an act of observation, the observer and observed system interact when the observer U plus M form a single quantum mechanical isolated system is described by the wave function W. The observers cannot detach themselves from what they observe. The combination of M and U to make W is not unique. There are multiple ways the eigenstates of M can combine with different states of U to make W. If H_M and H_U are the Hilbert Spaces for M and U then the Hilbert space for W is $H_M \otimes H_U$. The symbol $\otimes$ representing what is called the tensor product. We need not get into tensor product mathematics in any greater depth for present purposes. If the complete orthonormal sets for M and U are m_i and u_j respectively, the state W can be written as the superposition:

$$W = \sum_{i,j} (a_{i,j} \cdot m_i \cdot u_j)$$

where the $\sum_{i,j} ()$ operator is summation of whatever is inside the () by stepping the induces i and j over their ranges to get the items to be summed. The states M and U in Everett's words do not possess anything like definite states independent of one another unless all the constants $a_{i,j}$ but one are 0. However, if we choose a definite eigenstate say m_k as the state of M, then a definite <u>relative</u> state of U exists. It is:

$$U_k = N_k \cdot \sum_j a_{k,j} \cdot u_j.$$

N_k is a normalizing constant. Thus, each eigenstate of M leaves U in a different state when U interacts with M.

In what follows we are using the word *real* as an adjective, not a noun describing an element of a complex number. All of these states are equally real but they do not result in common states of awareness in U. M has been expanded in the orthogonal eigenstates of the observing operator which is a property of U. The eigenstates m_k i=1,2,3.. are orthogonal – you can't make one out of any linear combination of the others. They are in this sense mutually exclusive in the awareness states they produce in U which we can label as $U_1, U_2, U_3, \ldots$ The indexes may have a finite limit, but may also be infinite. Every eigenstate of M leaves U in a different state of awareness. Everett, without committing to an interpretation of reality, made the point that there is no reason to believe that any one of these states is more real than another.

We do make such a commitment here: If we leave the awareness of what was observed in computer storage and never look at it, we can do good science without thinking about reality. The merit of putting you at the terminal of awareness is that you can be aware that the U state of awareness in which the observed eigenstate of M left you was experienced in consciousness. This state of consciousness will not bind in awareness with any other although all, and they may be infinite in number, are equally real. For example one may be the conscious experience of the photon being reflected by a beam splitter and not being transmitted. Equally real is the experience of the photon being transmitted by the beam splitter and not reflected. This is impossible if reality is the "physical world" of classical mechanics and intuition. There is no need to hypothesize multiple conscious selves: Both happen to the same conscious self. ***Tat tvam asi.*** They are mutually exclusive paths in awareness which are real because they are consciously experienced by you, and not because you are a thing in an imagined physical world. If you believe you are a denizen of the physical world, in your imagination you have to collapse all the other eigenstates of which you are not aware into non-reality. Heisenberg et al did this. Everett pointed out that doing so is needless metaphysics.

Everett called this the **Relative Sates Formulation** of the quantum mechanics. Not Everett, but others called it the **The Many Worlds** interpretation, which is unfortunate for the purposes of this epistle, because it steps back into the very metaphysics that needs to be purged. This interpretation spawns multiple and generally infinite numbers of physical universes with consequent infinite multiplication of their energy content. Interpreting reality as consciousness does not incur this multiplication because consciousness does not exist. No one has ever measured a joule of consciousness. This is not metaphysical or mystical theory. Remember your last bad headache. That was not a mystical belief. Consciousness is not a phenomenon. It is not inside energy nor as von Neumann imagined inside the presumed physical world or it would be observable. Nor is it a spin-off of higher brain function. Energy is property of awareness and it is conserved in every consciously experienced history. Before we are tempted to theorize about consciousness, we need to recite thrice the reality catechism given us by Father Erwin:

> *Consciousness cannot be accounted for in physical terms*
> *For consciousness is absolutely fundamental*
> *It cannot be accounted for in terms of anything else*

Although all the U states are equally real, they do not bind in awareness because of the orthogonality of the eigenstates of the observation. Causality is rigorous – there is no need for an "Uncertainty Principle". Because all of the caused states U_i are equally real, there is no need for any to "collapse" all the others because you find yourself in one of them. The Copenhagen Interpretation reduces to pure metaphysics because it is predicated on the ultimate reality of the physical world and cannot tolerate multiple inconsistent copies. The non-commutativity of complementary operators remains valid, but can no longer be used as a crutch for the Copenhagen Interpretation to limp into metaphysical realms of randomness and superluminal causality.

In general the U will form a continuum: The reality of consciousness is thus infinitely vaster than awareness. These conscious states do not bind in awareness merely because a corpus callosum is missing between them. Their binding is impossible because they are caused by orthogonal eigenstates of the observation. For example a photon being transmitted by beam splitter and also being reflected by it are equally real in consciousness. We can only experience both of them in common awareness if both cause the same single event. Quantum math gives a simple rule for this case: The amplitude for this event to happen is the sum of the amplitudes for the event to be caused by each eigenstate independently. If we put photocells in each beam path, the conscious experience of seeing the photon with its full $h\nu$ energy in each path will be equally real: It is impossible for them to bind in a common state of awareness or realty would be inconsistent within that state and the law of conservation of energy would be violated. There is no inconsistency if we recognize that the law of conservation of energy, and other conservation laws, are in the mathematical structure of awareness and do not need one to believe that consciousness exists. If it makes one more comfortable, one can invert von Neumann's nesting and believe instead: The physical world is in consciousness: You will always consciously experience being in one internally consistent physical world or another. The ultimate reality of what you experience is consciousness and not the physical world.

If instead we believe in the ultimate reality of the physical world and try to preserve the equality of reality of the photon both being transmitted and reflected, which are forced to believe to account for interference, we are forced by each reality to collapse the others: This impossibility in turn forces us to believe that one reality or the other was selected by random ordination at the moment of observation; any reality not selected being instantly collapsed into non-existence. In the EPR experiment we cannot decide which observer was the cause of random ordination: If the observations are separated by times less than the transit time of light between them, the Special Theory disallows us to decide which happened first. This metaphysical mind warp can be avoided by getting rid of its cause: Belief in the ultimate reality of the physical world. One does not have to worry about what one needs to believe instead: Consciousness will be just as real no matter what one believes, including unmitigated nonsense.

With each state of awareness there is an associated amplitude and hence Born probability which is interpreted as a density or weight in awareness. All consciousness states pointed at by the unit eigenvectors of Hilbert space are equally real. They are experienced by the universal conscious self, if one needs to think of consciousness as self-hood. If so, such self-hood is not denumerable, or one returns to the paradoxes we are trying to eliminate. Bob and Alice are the same conscious self regardless of the similarity or difference between them in awareness.

This is the point of departure of physical world metaphysics and consciousness as ultimate reality. The physical world meta-physicist believes that if he is conscious of being in state U_3, say, then none of the others can be real. They have to be collapsed as soon as he is aware of being in U_3, because he cannot believe the physical world can be in contradictory states simultaneously. Otherwise, against his creed, he must abandon belief in the ultimate reality of the unitary physical world. He must believe U_3 is the only reality selected to be real by the toss of God's dice. If not, why didn't God select U_2 instead? The others God must have "collapsed" or you would be aware of them. Or, the physical world meta-physicist can accept Everett and still hang on to the physical world by believing multiple almost identical physical universes spring out of each quantum mechanical event. These pseudo problems do not arise if one accepts consciousness as reality because the states caused by different eigenstates of M do not bind in awareness. <u>We can think of the unit vectors of Hilbert space as pointers to states of consciousness and the normalized state vector as a list of amplitudes in awareness.</u>

Within this framework, no matter what path one follows. the conservation and symmetry laws will be obeyed, so the events of

167

physical experience will always unfold in time in a three dimensional space in an apparent physical world. If you look at the time on your wrist watch, what happens is only real in consciousness. For the most part you will imagine your watch as a physical object in 3D space because your neurology is programmed to see it that way without requiring you to be aware of the reality of your experience in consciousness. You will not experience your watch as a quantum mechanical object, which it really is, because your neurology is not programmed to do so even it you can comprehend a quantum mechanical world intellectually. Even if you can, the history of the past ninety years shows that this ability will not deliver you from belief in the ultimate reality of the physical world. The schizophrenia of the Copenhagen faction, while wallowing in "which path" mystery mongering, lies in its acceptance of the simultaneous reality of the plurality of eigenstates of the observed system when needed to account for interference. Here, it has no choice.

How can we escape these metaphysical contortions?

It is very important to escape from the following trap: When we observe the polarization state of a photon or the spin of and electron, we believe we are looking at what state it is in, and we only change to the extent of information entering our minds describing that state. This is the physical world model of reality. In reality, the eigenstates we observe belong to you, the observer: The particle will leave **you** in eigenstates of the observing operator **you** chose to make the observation: See the above math for details.

We adopt Everett's theory but do something he did not do forcefully: We assert reality as consciousness; we do not view reality as some imagined physical world. We can, in principal, avoid talking about reality all together: The reality of consciousness is not scientifically demonstrable, even though the proof of its reality given to you is incontrovertible. The danger of leaving an unoccupied place holder for reality is that meta-physicists will start filling it with illusions supporting their beliefs, while being convinced they are talking physics. This is exactly what happened to Everett's formulation. Historically none of the founding fathers of the quantum mechanics were immune, with the possible exceptions of Schrödinger and de Broglie.

It is very hard to stop believing in the ultimate reality of the physical world because it is the model used by intuition: For any situation where we can use 0 for Planck's constant and ∞ for the velocity of light, this model is very, very accurate in awareness in predicting experience in consciousness which, for the most part, we simply ignore: The model is never experienced except in consciousness without which experience would not be real. We simply create reality in the image of the physical world. I must repeat the disclaimer: I am only using the words *consciousness, awareness, existence* and *reality* as defined above. This creation has not always held sway in human awareness. American Indians attributed reality to the Great Spirit. The physical world model first took on a precise understanding in awareness with Euclidean geometry in ancient times. A thousand years later, it gave a precise understanding of the laws of refraction and reflection of light. The laws of inertial and gravitational dynamics of Newton married to Euclid's geometry gave us mathematical machinery to represent experience of nature to one part in 100,000,000. Finally, two thousand years after Euclid, Maxwell's electromagnetic theory encompassed all that was known of electric and magnetic force, and married Newton's laws. A child was born: Ultrah Violet Newton Maxwell Catastrophe. Her cataclysmic apparition is still with us.

Understanding quantum cause and effect

We expel from our minds belief in reality of the physical world. Do not even use it as a crutch – one only stumbles back into mind boggling paradox. The existential properties of elementary particles which are measurable such as number, charge, spin, position relative to a distance scale at a time indicated by a clock are properties of consciousness and do not constitute a basis for believing that "the physical world" is real because such mathematical properties exist.

We only recognize the reality of consciousness. The simultaneous reality of inconsistent conscious states is not a problem as long as they do not bind in awareness. Each corresponds to an internally consistent state of awareness. Inconsistency across conscious states will always be true if the state vector of the observed system is a linear superposition of multiple eigenstates of the observation. The eigenstates are orthogonal: None can be represented by any mix of the others.

Strict causality is restored to the quantum mechanics. Causes are the eigenstates of the operator representing measurement of the system observed giving non-zero amplitudes to be in each eigenstate. All eigenstates are equally real even if the corresponding amplitude magnitudes are very small.

Effects are states of awareness of the observer caused by the eigenstates of the observed system. Reality is the conscious experience of these states of awareness. When this is a plurality, these states of consciousness do not bind in awareness because of the orthogonality of originating eigenstates.

Reality is not created in the image of the states of awareness so there is no inconsistency in the reality of states that are inconsistent. For example two eigenstates of a photon may be real that are inconsistent in awareness like

1. transmitted by a beam-splitter and not reflected
2. reflected by a beam-splitter and not transmitted.

When there are multiple observers, we recognize only one conscious self as the conscious observer. *Tat tvam asi.* Consciousness may be distributed in non-binding awareness of human observers or even non-binding awareness in a single brain.

The observer need not have a brain. A Pentium II will do. So will a piece of wire. So will a free electron: The observer then is the observed particle: the particle itself will constitute its own state of awareness even if it exists in multiple eigenstates. Because we do not have to discuss reality, we do not care as humans what it "feels like" to be a Pentium II. Or a piece of wire. Or an electron.

Interference describes the effect caused by multiple eigenstates of the observed system, having an associated amplitude which is the sum of each amplitudes to have been caused by each eigenstate, or the integral over these states where they are a continuum.

While inconsistent eigenstates may be real (consciousness) where they do not bind in awareness, interference cannot be real if any of its contributing eigenstates also exist in awareness. This is equivalent to the statement that interference is destroyed by the existence of "which path" information, or restored by the destruction of "which path" information.

Look again at this apparatus discussed earlier, which is just the first beam-splitter BS of Mandel's interferometer:

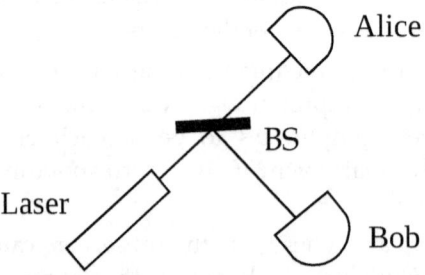

We have put photometers in each path. One, top = Alice, sees transmitted photons. One, bottom = Bob, sees reflected photons. After each photon passes the beam splitter, it will be in two eigenstates of the beam-splitter, M_{alice} and M_{bob}. Had there been no obstruction in the paths after the beam-splitter, the paths could extend across intergalactic space for billions of light years and the photon would remain in eigenstates M_{alice} and M_{bob}. But we have placed photocells in each path and you are observing them. W is the state vector of above apparatus, the photocells, you and as much of the cosmic universe as we wish to consider an isolated system, which can arise in two ways after the photon is passes the beam splitter:

- U_{alice} : aware that the photon was transmitted and not reflected
- U_{bob} : aware that the photon was reflected and not transmitted

Both of these states of awareness of U will be experienced by as real states of consciousness that do not bind in awareness because of the orthogonality of their originating eigenstates. There is only one *you* (tat tvam asi) experiencing these conscious states but they do not bind in awareness. Both Alice and Bob are the same conscious self but their brains are two different mechanisms of awareness. The question "which path did the photon take is meaningless". It took both paths whether or not these paths are destined to interfere. This view is not consistent with one seeing the game playing out in the Physical Universe unless you are willing to believe that as the photon passed the beam-splitter the universe split into two. Otherwise one has to understand that the word *you* refers to consciousness as ultimately real.

If we use Dirac's notation and call

|R> the reflected eigenstate of the photon and orthogonal to it the

|T> the transmitted eigenstate

The photon is in a linear superposition of these states after passing the beam splitter:

$$(1/\sqrt{2})|R> + (1/\sqrt{2})|T>$$

assuming, for simplicity, the equal Born weights.

Once Alice is aware she is in a state of awareness caused by |T> she can set her amplitude to be in |T> to 1 and her amplitude to be in |R> to 0. This does not "collapse" |R>. |T> also causes Bob to be in the same state of awareness for future calculations that he did not see the photon, but Alice did.

The physical world metaphysicist cannot countenance the following truth simultaneous with the above. Copy and paste the above sentence interchanging Alice and Bob; T and R:

Once Bob is aware he is in a state of awareness caused by |R> he can set his amplitude to be in |R> to 1 and his amplitude to be in |T> to 0. This does not "collapse" |T>. |R> also causes Alice to be in the same state of awareness for future calculations that she did not see the photon, but Bob did.

Everett pointed out that both of the above states of awareness in italics are equally real, but both will never be experienced in the same conscious state. They were caused by orthogonal eigenstates. It does not matter how far apart Bob and Alice are from one another, or whether or not the exchange notes, but it they do, in both cases in italics above, they will always agree that if one saw the photon the other did not. Once one recognizes that conscious is real and the physical world is not, the mystery mongering about superluminal eigenstate collapse evaporates into purple smoke and blows away.

However complicated Alice and Bob's brains may be in consciously experiencing awareness, if the photon is seen by Alice it will not be seen by Bob. Conversely both of these states of consciousness caused by the superposition state of the photon are equally real because these states of consciousness do not bind in awareness. Everett showed that Alice and Bob share the same awareness. This is impossible in the physical world which cannot exist in two conflicting states of reality. But it can be true of consciousness because it does not exist. The states of awareness do exist but do not conflict because they do not bind the conscious states into a single state which is internally inconsistent.

To understand how this can be, think of the example given earlier of what happened when you were struck by lightening and suffered amnesia: Your memories before the accident and those after it are disjoint. However in the case of amnesia, the eventual union of the two states of awareness is not impossible. In the quantum mechanical case it is impossible unless the apparatus is changed to make both M_{alice} and M_{bob} cause the same event to happen: interference. If it were possible, as will be discussed later, we could remember the future.

This interpretation has been called the Many Worlds Interpretation of the quantum mechanics attributed to Everett. Its merit is that it discards the metaphysical uncertainty principle without discarding the underlying physics of non-commuting operators. Its demerit is its invitation to see reality as an ultimately real physical world created in the image of $h=0; c=\infty$, which not only leaves you in two disjoint states of consciousness that do not bind in awareness but also creates two of your bodies, two Eiffel Towers, two Grand Canyons, two Mt. Everests and two

Andromeda Nebulae.

I believe, however, Everett is done an injustice by attributing the Many Worlds interpretation to him. He recognized the need to keep open the door to consciousness as ultimate reality, by pointing out that his analysis does no violence to the doctrine of psycho-physical parallelism. He did not use the word *world*.

Because consciousness, although real, does not exist, it takes no more energy for two conscious states to be real instead of one. All of the conservation laws of physics remain intact no matter which eigenstate of M causes them. Instead of being different physical universes, they are different histories in consciousness, all equally real. Each history is a consciously experienced awareness of which eigenstate of the observed system caused it and will contain no awareness of any other. Conscious experience will always be seen to happen in a physical world in which the conservation laws of physics hold. This is what the doctrine of psycho-physical parallelism says. The physical world is thus a property of consciousness. If you try to turn this inside out and believe that the physical world is ultimate reality and consciousness is some kind of phenomenon inside physical reality, you are back in the paradox of multiple physical universes with multiple energy content. If you reject Everett and follow von Neumann, you also have to let go of causality, because the same precursor event can have different consequences chosen apparently randomly. Finding yourself in one of them, you need to get rid of the ones you didn't experience by "collapsing" them. And explaining to yourself how the one in many, if not infinitely many, was chosen to be real by saying the "Uncertainty Principle" chose it . And explaining why when you push the button again the "Chosen World" was not chosen again. Except sometimes. This is the metaphysical turmoil that comes from trying to hang on to the $h=0; c=\infty$ world by refusing to believe that it is not real.

Believing in the reality of consciousness is not a hard step to take. The hard thing to do is letting go of the belief in the ultimate reality of the physical world. One grabs at straws: "But electrons and protons are real" one cries. Indeed they are: They are the real properties of consciousness.

They are not real because they are elements of a delusion: The Physical World. The photon in the above experiment is equally real when reflected by the beam splitter as it is transmitted by the beam splitter. One cannot have two physical universes to accommodate both of them without absurdity, but both can be accommodated by no physical universe at all. Invoking the physical universe to explain

173

electrons, quarks and photons is like me telling you because houses can be haunted, I believe in ghosts. When you tell me you do not believe in ghosts, I ask you: "You mean you do not believe houses exist?" The existence of electrons, quarks and photons does not mean that the physical world is anything but a delusion, to be thrown out like dirty bathwater without throwing out baby electrons.

Suppose you repeat the above Alice/Bob experiment 3 times. In each performance in each state you will remember your precursor at the previous node, and all its precursors:

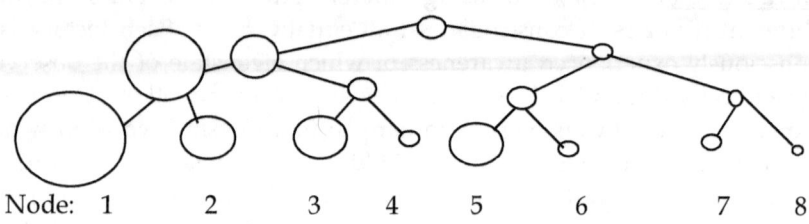

Node: 1   2   3   4   5   6   7   8

There are 8 final states all equally real. The path to each terminal is a different history of events in consciousness. At every node in the above diagram <u>exactly the same thing happens and it is very simple and easy to understand:</u> Nature takes Yogi Berra's advice: When nature comes to a fork in the road, it takes it! The photon enters two **equally** real states: It is both transmitted and reflected by the beam splitter, after which it exists in two equally real states in different regions of space. The quantum mechanics math tells you **exactly** what you are going to experience in consciousness. And that is **exactly** what happens to **you** in **consciousness.** There is no **uncertainty** about it. Amplitudes are represented by the size of the ball. We are supposing here that the beam-splitter is not symmetric: The square of the amplitude to go left is twice that of going right in this example. If the experiment is repeated many times, final node 1 will be experienced most often and final node 8 least often. There is only one conscious self at all of the terminals in the above diagram (tat tvam asi) but each is a different self in awareness who is only aware in memory of the single path that brought them to that terminal. The conservation laws of physics are obeyed along all of the paths so no matter what terminal you end up in it will always seem to be "in the physical world". Suppose you send three photons through the apparatus but find yourself in node 5 in the above diagram. If you ask "But why node 5?". **The answer is node 5 is real in consciousness and it had to be you. Tat tvam asi. This is not a random choice. It was absolutely certain that you had to be in node 5, and hey, you are there.**

Suppose, before doing this, you did not <u>ever</u> want to be in node 5. In that case do not do this experiment. Suppose you said "but I <u>want</u> to be in any node but 5". Then change the apparatus so that the amplitude to be in node 5 is 0. Put it in the center of a dark interference fringe.

If you repeat the same experiment many times over and tally the scores in each of the terminal nodes of the above diagram, you will find that the tallies for any concatenation of remembered events converge on the probability predictions of quantum mechanics, justifying Born's statistical prediction that the distribution will be the absolute squares of the amplitudes to be in each node. But this is true of every terminal node in the above diagram. **Yet, in Everett's formulation, you can predict with certainty exactly what happens in consciousness to you, tat tvam asi, every time you transmit another photon. Each state of awareness will exclude all the others and will begin a new history in consciousness. If you are not happy with what you get, you are doing the wrong experiment. This has no historical precedent, but never in history before the quantum mechanics has it been necessary to make the distinction between reality in consciousness and experience in awareness.** When you are conscious of being in a state of low amplitude the reality of consciousness is the same as that of being in a state of high amplitude. Reality has nothing to do with the amplitude of being in some state. You will experience it if you are there and will not if you are not there. The conscious reality of seeing the Mona Lisa has nothing to do with how many people are looking at it. Or how many eyes you are using, or which eye you are using, modulo body parity and asymmetry.

Dirac pointed out that eigenvectors are directions in Hilbert space. Their lengths do not point out how much reality lies in the direction they point. They point at states of consciousness which are all equally real. They point at states of consciousness for which the corresponding awareness has an amplitude. If we extend the above tree indefinitely and look back along the only path we can remember, the states with greatest amplitude will occur most frequently in awareness and conversely. When a state with very small amplitude is experienced, it will be just as real as any other.

Dirac also pointed out that the magnitude of vectors is meaningless, only their ratio matters. The amplitude of a chain of events to happen is the product of the amplitudes for each link. Thus all past history collapses into a single number. It is traditional to set the amplitude to 1 at the start of the chain we are investigating, because we are certain we are there, so the computed probability of being there is 1. However, does God have to do this? If God started the amplitude at 1 at the Big Bang, the amplitude to be anywhere now will be very, very, very small. But the conscious experience of being there will be just as real as being anywhere else. Do not worry about God running out of small numbers. Below any number however small there are always infinitely many that are even smaller. Dirac said all this in fewer words.

Coherence time. A photon in "physical reality" is a mathematical point with a definite position in space at each instant of time. This is an illusory view of reality. The wave function of a particle and a photon in particular extends over space and time. This means that anywhere in this space time volume it has non-zero amplitude. To know both its wavelength and when it seen, we have to make two measurements. In an interferometer, the wavelength, and hence the photon energy, must be exactly defined or we will not have interference fringes, but the photon may exist in states that are meters apart. We cannot put a tight window on when the photon made its journey or the photon will exist in a continuum of eigenstates of energy and hence a range wavelength that will prevent the interferometer from working. The size of this time window is what has been called the coherence time above. The photon has negligible amplitude to be outside this window but may have finite amplitudes anywhere inside it. This is more than analogous to putting on a photon coming from a star in the space window of the astronomer's telescope aperture, discussed in the chapter *Uncertainty Principle*.

It may seem that we can trick the photon into telling us when it made its journey. Poke a photocell into the beam at some definite point in time in the coherence time window and quickly pull it out. If we catch the photon we will know when it made its journey. But in this case it will never get to the screen. We say that does not matter because we are willing to throw that photon away, but we will know something we didn't know before – when when other photons did **not** make their journey, since they went by earlier or later. So we can do it again at a different time in the coherence window and sacrifice the photons we caught, but get a good picture of when the others did get through. This is a delusion. Doing it will systematically destroy the interference pattern of arriving photons that we believe could not have known the window was blocked.

Putting obstructions in the time window destroys the pattern of arrival of photons, even though none were stopped by the obstruction, because we tighten the time window for photons that got through. When we block windows in time, we are setting to zero amplitudes that would cancel others at energies we do not want the photon to have, lest interference be corrupted. We loosen the energy or wavelength window which degrades interference in an interferometer. The photon is interfering with itself: The Fourier transform at work.

Astronomical telescope: An anti-interferometer

In the chapter titled **Uncertainty Principle** we showed that, with the exception of the sun, the image a telescope makes of a star is a pure interference pattern: For economic reasons, we have to throw away hundreds of square light years of the wave function wave-front of this "particle" called a photon. Metaphysical question: How in heck can a particle be spread over 100 square light years?: Dirac got a Nobel Prize for teaching us the physics that tells us how: The location of **every point on that 100 sq ly surface is an eigenstate** of your observation of photon position and they are all orthogonal to one another. If you look at the whole surface, you certainly will see the photon. If you find yourself saying "but you couldn't predict where I would see it", please reread the sentence above in **bold** type. It includes the point you saw. Its being there is real and it had to happen to you. *Tat tvam asi.*

Putting obstructions in the photon's time window is also true of spacial obstruction earlier, which is easier to understand. A good way is to look at the quantum interference effects in an astronomical telescope: a device that exhibits quantum interference, which is exactly what astronomers do not want but it to do, but have to live with.

If we took a photograph of the sun at the distance of the nearest star using a perfect telescope with a focal length of 50 meters, the sun's image would be as small as a microbe. Such a telescope would have to have an objective lens or mirror many kilometers diameter to see details like sunspots or the transit of a planet. If each photon fills an enormous space across the line of sight, we can know the photon's momentum across the line of sight almost exactly, so all the observed photons would aim exactly at the microbe sized spot. But the Hubble space telescope which gives the best images of stars free of atmospheric degradation would have an image size a hundred times larger. Telescopes like those used by amateurs would make images thousands of times larger. These are called diffraction patterns which obscure all the details we would like to see. They were discussed in the **Uncertainty Principle** chapter. Nonetheless, the telescope is a clever design which makes the amplitudes for each photon reinforce where we want the image to be and cancel where we do not want it to be. A good telescope can be described as the best anti-interferometer one can design for the money. The loss of resolution is caused by our throwing away most of the wave function, all parts of which are non-zero, are real. Economically, getting more of the wave function rapidly gets more expensive.

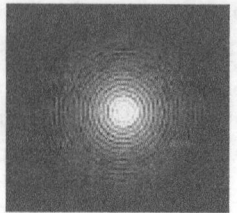 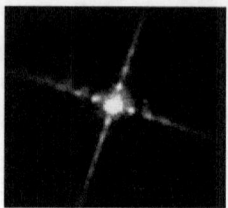

The images shown above are star images seen at high magnification through different telescopes (probably not the same star). The actual star is almost a mathematical point at the center of each image. The image on the left is typical of the image seen in Maksutov Cassegrain telescopes which have a secondary mirror in the light path that is deposited on the surface of a transparent lens which fills the aperture of the telescope. The second picture is typical of a Cassegrain in which the secondary mirror is supported by metal vanes called a spider. The streaks in this picture are caused by the spider. It is tempting to say: "Light reflected off the vanes. They should have been painted black". They were painted black and absorbed any photon that hit them. It is tempting to ask: "Well, if they didn't bounce off the spider what <u>did</u> they come from?" The metaphysicist inside us does not want to hear the truth: They came directly from the star to the image detector missing the spider vanes vanes. If you find yourself asking the question: "So how could they have known the spider was even there if they missed it?" you are caught in the metaphysical Physical Reality trap. Don't feel bad: Einstein was too!

The patterns are represented for the arrival of <u>each</u> photon by the sums of amplitudes for all available paths it could have taken through the telescope optics from the star to the image plane. But the paths through the spider which give finite amplitudes at the image plane were the spider <u>not</u> there are exactly what is needed <u>to cancel the spikes</u> by having the amplitudes there add to 0. In this sense a good telescope is an interferometer where the amplitudes sum to 0 where we do <u>not</u> want the photon to go. If the spider <u>is</u> there, it prevents amplitude cancellation and that causes the streaks. The streaks are caused by <u>each</u> photon taking an infinitude of paths that did <u>not</u> hit the spider vanes but which had a non-zero amplitude to arrive there that would have been canceled to zero had the spider not been there. Ergo the name anti-interferometer: It does the best job for the money of making amplitudes add to zero where we do not want the photon to be observed. The astronomer's choice is the lesser of evils: If the spider is not there, the secondary mirror falls on the floor smashing it, or, worse yet on the primary mirror, smashing it too. Don't laugh. A mechanic dropped a hammer on the Mt. Wilson 100" primary.

When we say that what happens quantum mechanically depends on what you **know**, we can say here that the streaks caused by the spider are caused by our knowing the photon could not have gotten to the screen by any route passing through the opaque spider material. The spider actually stops very little light because the vanes are thin black sheet metal with surfaces parallel to the light path. That's why the streaks are only seen with very bright stars: They make for pretty pictures like this one from the Hubble telescope:

But they are the bane of astronomers.

In the real quantum world, what happens is governed by amplitudes everywhere and every-when in space and time where the amplitude is not zero and that can include large swaths of space and time. To get the amplitude that the particle will be at some point P in space at some point in time we have to calculate the amplitude for a classical particle to have made that journey taking into account everything it encounters along the way in the classical world at the time it was there. It has been shown experimentally that changing things somewhere at a time the classical particle would not be there has no effect. After doing this for every possible path, we add the amplitudes together to get the amplitude for the particle to be at P at any time. This addition generally involves the calculus operation called *integration.* Feynman described this by saying the particle "smells out" every path. It does this both in space and time. To be sure, when the particle is observed, it is seen to be at a particular point in space at a particular point in time. Here the quantum mechanics and classical physics do not disagree.

Thus <u>every single photon</u> knows exactly where every atom of the spider is located. If no atom blocks a particular path, the photon will have some amplitude to reach the point where the path hits camera image plane. If some spider atom gets in the photon's way, that amplitude will be zero. That's a heckuva lot of intelligence for one little photon. Sorry, folks, this is **not** "physical world" physics.

With Everett's interpretation where we observe each eigenstate of the particle as a real state of consciousness in which all eigenstates are equally real, our awareness is its only being in one point of space-time. If we do not have a theory that represents the data, we can repeat the experiment many times and analyze the results statistically to infer the wave function. When we look at the diffraction patterns of stars in the above pictures, we are looking at the reality of consciousness reconstructed from multiple states of recorded awareness which bind in a single state of awareness because we remember the past states. This is how astronomers were able to get the exact locations to $1/_{100}$ arc-second of fundamental stars using meridian circle telescopes with a resolving power often worse than an arc-second. Before a star could be considered fundamental for star charting, it had to be examined using a much larger telescope to make sure it was not double, or it would not classify as a single fixed beacon that did not wobble.

What Hugh Everett gave us:

It may seem to be outlandish to make the following claim, but I do so in all sincerity: Of all the scholars of the quantum mechanics in its nigh century history, Hugh Everett is the only one who has truly understood the awesome message its mathematics teaches us. He showed that the representation of experience by the quantum mechanics can be understood completely as following from the equation:

$$\partial\psi/\partial t = A\psi$$

where **A** is a linear operator, ψ is the state vector of an isolated system and $\partial/\partial t$ is the time rate of change operator. This equation is completely deterministic: Nothing random; no collapsing reality, is needed to understand its consequence. It alone explains everything we experience.

He made no appeal to preconception of what is real, but if it is understood to be consciousness, this equation embodies a richness infinitely vaster than the physical universe. However basic the math, the magic is in the linear operator **A** which may encompass anything, even the entire cosmic universe. Why is it so hard for others to understand? Everett was a smart mathematician, but so are they. Before being physicists, they are meta-physicists. Their whole lives long reality has never been manifest to them except in consciousness, but they will push it aside and look for the physical world beyond.

Of course there are, and have been in the past, large numbers of scholars who understand quantum mathematics very well. They fall into three groups depending on their metaphysical persuasion.

- The first, like Feynman, do not even acknowledge Everett's theory of awareness. They accept as fact the metaphysical artifices of random and superluminal causality. In Feynman's words: *Yes! physics has given up. We do not know how to predict what would happen in a given circumstance, and we believe now that it is impossible—that the only thing that can be predicted is the probability of different events. It must be recognized that this is a retrenchment in our earlier ideal of understanding nature. It may be a backward step, but no one has seen a way to avoid it.* They do this, I suggest, because they cannot abandon faith in the ultimate reality of the physical world, whether this faith is grounded in the intellect or gut level intuition.

- The second, like DeWit, do acknowledge Everett's theory of awareness, but still hang onto the physical world by believing that each observation of a system whose state vector is not an eigenstate of the observation splits the physical universe into many, and possibly infinitely many, copies of itself each containing awareness of one eigenstate of the linear superposition of eigenstates of the observation that make up the state vector. This absurdity, seen as multiple physical universes, is so extreme that it seems to have driven most second group believers back to the first group.

- A third group: Einstein, Podolsky and Rosen, appealing to faith in "objective reality", believed that the quantum math was not a complete theory. After Einstein's death, it was shown by Bell that what they proposed was experimentally testable. Aspect's experiment showed the EPR proposal to be wrong.

Everett showed us that quantum math based on Schrödinger's equation teaches us all we need to know about the structure of awareness, without interpretation. I do not believe he can be faulted justifiably for not making an interpretation of reality. Whether this was for reason of caution or wisdom, I do not know, but he really didn't have to: The reality of consciousness is demonstrated to each of us every waking second of our lives and this reality does not play second fiddle to what anyone believes.

When one considers the simplicity that Everett's work brings to the quantum mechanics on may wonder that anyone who believes otherwise is stupid. Einstein believed otherwise; if one thinks he was stupid, good luck convincing the rest of us. But how could he not have seen the truth? The past sixty years show that Everett's work not to have made its mark follows from his reticence to address reality seen in the context of that work.

The Copenhagen Interpretation put the observer firmly in the theoretical homeland of objective reality: The Classical Realm. Einstein believed in objective reality and re-configured the quantum mechanics to make it comply with this preconception. His proposal ended in failure.

Consider this example of the role of objective reality: You receive a citation from the department of motor vehicles that you ran a red light. You don't believe it and go to court. Submitted into evidence is a photograph taken by an automatic camera on the traffic light standard of you car entering the intersection. There is no evidence in consciousness of this but there is evidence in awareness recorded by instruments. You admit to driving through the intersection but entering it when the light was yellow. The police testify that the camera is timed to take the picture a second after the light turned red. Your lawyer submits evidence that the camera can be triggered by power glitches. The police agree but say it rarely happens. You are exonerated and awarded legal costs. The traffic department changes the camera so that the car and the traffic lights, in future cases, will appear in the same photograph. Most of us agree that the game played out in objective reality, not is subjective imagery. We would not like to be told it is stupid to do so. However, the whole game was played out in consciousness by those in court who looked through it at the objective world of cameras and photographs. These things play the game following the rules of the classical realm. The conclusions would not be altered if we used quantum mechanics. Cars and cameras are made of electrons, quarks and photons which are not elements of the classical realm and does not play by its rules: The best understanding we have of the game they play is the quantum mathematics and theories of relativity.

More that half a century after the birth of the quantum mechanics, and almost three decades after Einstein died, Aspect experimentally showed Einstein to be wrong: Bell had shown that Einstein's version made mathematically different predictions than those of pure quantum math. Thereafter, the classical realm no longer was considered to have any valid basis.

Einstein died a year before Everett submitted his thesis. Had he lived long enough to read it, the future of quantum interpretation would have been very different for three reasons: It removes his:

1. objection to randomness in causality: God plays dice;
2. objection to superluminal causality: Spooky action at a distance;
3. need to make quantum math complete.

We will never know whether these things would have broken Einstein's death-grip on objective reality.

Why has this metaphysical turmoil lasted a century? I think it is for this reason: Feynman's words show that he knew was in a frying pan: *"Yes! physics has given up. We do not know how to predict what would happen in a given circumstance, and we believe now that it is impossible"*. His words show that he could see no place to jump: *". . . but no one has seen a way to avoid it"* All scientists worthy of the name know by now: One cannot infer natural law by philosophical contemplation of what one thinks is "real". However, the lesson this century of the quantum mechanics is: Physicists have shown that they must have a place to ground their beliefs in <u>something</u> they feel the mathematics of physics is talking about. Reality must be something. If necessary, they will invent it or grab at anything in reach. Everett gave them no place to jump, so either they:

• stayed in the frying pan with Feynman even at the price of believing in metaphysical indeterminacy and collapsing realities to keep their haven consistent. If asked about the reality of both paths of an interferometer, just bury your head in the sand by <u>not</u> asking *'But how can it be like that?'*, at the price of going down the drain; or they

• stayed in the physical world like DeWit, even at the price of inventing infinitely many of them; or they

• like EPR, jumped into what they saw as "objective reality" and ended up in the fire.

The Einstein, Podolsky, Rosen Experiment

In the light of Everett's formulation of the quantum mechanics to which we add the interpretation of reality as consciousness, we describe Feynman's version of the EPR experiment, as it is commonly called, originally done by Prof. Alain Aspect.

Quantum entanglement

If there are two <u>isolated</u> systems, S and O, for system and observer, and $S(s)$ is the wave-function of S and likewise $O(o)$ is the wave-function of O. s and o are the respective coordinates of the systems: s is the collection of coordinates of for each of the degrees of freedom of S and likewise o for the O system, then if there is some amplitude S for S to be in some state and there is some amplitude O for O to be in some other state, the amplitude for both states to be true is O.S. This is a law of the quantum mechanics. In other words the wave function $Q(o,s)$ of both systems is $S(s).O(o)$, whatever o and s may be. In this case it is impossible for the observer to look into the O state to find out anything about the S state.

However, if O has interacted with S at some time and is now isolated from S, examination of some states of O may reveal something about some of the states of S after the interaction ended. This has been defined to be *entanglement*. This should surprise no one. Back from Paris you are looking at your vacation photos and see one you took of the Eiffel Tower. You are seeing something of its present state provided it has not interacted with a third system like an asteroid in the mean time.

Quantum entanglement usually refers to a special relationship of two particles described by one wave-function: They were once together and now are apart and by a third system, namely you, by observing one you can tell the state of the other. The Copenhagen Interpretation, CI, has generated nearly a century of mystery mongering on the part of no lesser personages than Bohr and Einstein by seeing this entanglement as a paradox. This paradox arises because the particles may be in linear superposition of eigenstates, but CI says observing one particle will randomly collapse it into one of its eigenstates. But the other will collapse into the same eigenstate when any observation is made. Paradox: How does it know what happened to its mate? Einstein's hidden variables resolution of this problem was shown to be a failure when Bell demonstrated it predicted an inequality shown to be violated by Aspect's experiment, discussed earlier.

We will discuss the resolution of this paradox later in the chapter on Mandel's interferometer, when discussing the entanglement of two photons generated by parametric down conversion. Instead of describing Aspects experiment published 12 July 1982 in ***Physical Review Letters* 49**, 91 in which he demonstrated a violation of Bell's inequality by five standard deviations, debunking EPR, we describe a thought experiment of Feynman described in in 18-3 of Vol III, ***Lectures on Physics.***

Feynman generates two photons using the disintegration of positronium: an "atom" with a "nucleus" of one positron plus one electron, which are antiparticles of one another. Once such an "atom" is artificially created, its life is short before the antiparticles unite annihilating one another. In one decay mode, spits out two photons, 1 and 2. traveling in opposite directions. They can be right circularly polarized R or left circularly polarized L. The final state F is:

$$|F\rangle = |R_1 R_2\rangle - |L_1 L_2\rangle$$

which are the only two states for F that conserve angular momentum. Photon 1 goes to Bob in Reno and the other 2 to Alice on Proxima Centauri b four light years away. We can imagine the positronium decomposing into photons roughly half way between Reno and Proxima. According to Feynman it does not matter who sees their photon first – that is not an issue – if Bob sees R, Alice will see R. If Bob sees L, Alice will see L.

We ignore the fact that positronium photons are gamma rays so Bob and Alice can use calcite crystals to observe their photons in linear superposition of orthogonal polarization states. Called the ordinary ray x and the extraordinary ray y (Yes, that what physicists call these rays. Only ask historians why).

Now Feynman proves that if Alice sees x, Bob will see y. If Alice sees y, Bob will see x. He shows $\langle x_1 y_2 | F \rangle = \langle y_1 x_2 | F \rangle = i$ which is unit probability, but $\langle x_1 y_2 | F \rangle = \langle y_1 x_2 | F \rangle = 0$ = impossible.

Feynman says: There is no paradox here. What we observe and what the quantum mechanics predicts are the same thing: Alice and Bob's observations are correlated. However, saying "correlated" explains nothing: It only gives what is happening a name. Like saying they are "entangled" also explains nothing. He says that Nature describes the photons with interfering amplitudes and measuring destroys interference.

Nothing Feynman says in sub-chapter 18 of his volume III contradicts the conclusions of this epistle: There is complete agreement, including his words: *Nature apparently does not see the "paradox..."*. However, this statement is a tautology: **Nature never sees paradox.** Only people do. Feynman buries his head in the sand if he want you to believe that Feynman does not see the "paradox" by saying: '*...it is not a "paradox", but it is still very peculiar*'.

The paradox comes from nothing he says in sub-chapter 18, but in something he believes fervently, described in his words: *"Yes, physics has given up. We do not know how to predict what will happen in a given circumstance."* The choice of Alice seeing x is random as is the choice of Bob seeing y, yet they are correlated exactly. No, Feynman tells you: Do not ask how it can be like that or you will go down the drain. Advice in other words to bury your head in the sand too.

Feynman was one of the brightest scientist of history. If there is a simpler way to see things, why didn't he? For the reason, I suggest, that there is something he believed to the core of his being that he never questioned: The ultimate reality of the physical world.

The conscious experience of Alice seeing x and Bob seeing y is real. But the conscious experience of Alice seeing y and Bob seeing x is also equally real. These two conscious states do not bind in a single state of awareness. They are caused by orthogonal eigenstates of the decomposed positronium and exactly the same thing happens each time. There is no random selector deciding this will be real now and, maybe, that will be real next time. It will take at least two years for Bob and Alice to enter common states of awareness but the quantum mechanics promises they will be consistent when this happens. After N positronium atoms have decomposed there will be 2^N real pathways in consciously experienced awareness, but each will bind in only one pathway in awareness. If the physical world is real, these pathway can only be real in 2^N disjoint real physical worlds which we only can salvage with the Many Worlds Interpretation.

That consciousness is real, but does not exist, is is a simple way of saying that the physical world is a metaphysical delusion that is not needed. We are not abandoning our old physical reality home if we leave. We can always go back and look for our lost reading glasses in it, and we can go back to the flat earth when boating on the lake.

We can imagine that each photon is tagged by a radio message giving the photon number so they can compare notes years later. Bob and Alice send snail mail notes showing what they saw with each photon number, they will always agree exactly. It will always be true that x,y and y,x are real but these equally real experiences will never bind in common awareness.

Leonard Mandel

A professor at the University of Rochester in the 1990s, Leonard Mandel, built an ingenious interferometer to investigate certain metaphysical claims that had currency in physics.

These claims can be illustrated using a simpler interferometer than the one he used. We describe the first such instrument: Young's interferometer. Ignore the photocells blocking the light paths for the moment.

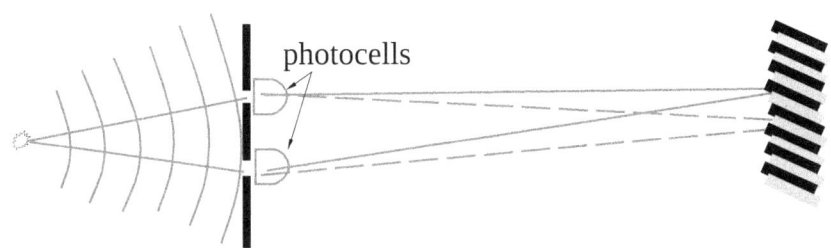

Light waves from a point source reach two slits in an opaque plate at the same time. The light that passes each of the slits will radiate away from the slits as though they were point sources, reaching the central bright fringe at the screen on the right at the same time, shown by the dashed lines. The light waves from each slit reinforces the waves from the other slit making the fringe, as it is called, bright. But the distances the waves have to travel to other points on the screen will not be the same. If the difference is a multiple of one wavelength there will be a bright fringe. If the difference is an odd number of half wavelengths the waves will cancel, making a dark fringe. An exact mathematical description of this wave interference and reinforcement was given by Huygens, a contemporary of Newton. It is true of any wave such as sound and water waves. This is the classical experiment by Young in 1803 that proved Newton's claim that light did not propagate as waves to be wrong. Although disputed at first, devices with multiple slits called diffraction gratings established the wave theory of the propagation of light beyond doubt and allowed the wavelength of light to be measured with very great accuracy. Electromagnetic waves make intuitive sense because the electric force fields of the waves cancel in the dark fringes resulting in no force on the electrons of the screen, be it a photographic plate or a photo-diode. If you and I are pushing a stalled car and you push from the rear and I push from the front our forces cancel and the car goes nowhere. These forces of the waves from both slits reinforce in the bright fringes moving electrons around and exposing the plate or creating electron-hole pairs in the diode. Now we are both pushing the car in the same direction. This wave cancellation and reinforcement is called *interference.*

There is no metaphysical mystery to this experiment if light is seen as an electromagnetic wave, some part of which goes through one slit; and some part of which goes through another.

The philosophical plot thickens when the light is so dim only one photon is making a journey at a time and you put photocells (gray in figure above) immediately behind each slit to see what slit the photon is going through. If you see the photon behind one slit, you will **never** see the same photon behind the other slit at the same time. So the photon is not splitting up with one piece going through one slit and another piece going through another. However, if you remove the photocells from behind the slits, the arrival of photons at the screen, on the average, will be the same as predicted for electromagnetic waves. <u>This is the philosophical humdinger that spawned the metaphysical debates of the quantum mechanics.</u> It would seem that a photon must go through one slit or go through the other. If it goes through only one slit why is the arrival rate affected by its going through the other at the same time? Why does it even know that the other slit exists? Photons will never arrive at the center of the dark fringes and will most often arrive in the bright fringes. If the screen is sensitive enough, you can detect the arrival of nearly every photon and they arrive one at a time.

Newton believed light propagated as corpuscles, and Young proved him wrong. But Planck was forced to accept light as photon particles or face an unacceptable disaster: The ultra-violet catastrophe. As described earlier in this tome, the case for photons was cinched by Einstein's explanation of the photoelectric effect.

Mandel's interferometer

For decades, physics students were taught that the interference is destroyed by messing with the particle on it way to the screen by trying to find out which path it was taking: **Until Prof. Leonard Mandel came along and said in effect: "I can find out exactly which path particle took without messing with it at all."**

Mandel built an interferometer similar to the Michelson-Morley device, which used a beam splinter like a half silvered mirror, but inserted barium borate crystals in each of the two beams. These crystals have the property that an ultra-violet (UV) photon going in comes out as two red photons. This effect is called *parametric down-conversion.*

We take the case where each red photon is exactly twice the wavelength as the UV photon. The red waves quantum mechanically are phase locked with the UV wave. Mandel recombined one red photon path from each crystal to make an interferometer and recombined the other red photon paths from each crystal to make another interferometer. *Phase locked* means that the sum of the quantum mechanical phases of the two red photons equals the phase of the originating UV photon, modulo an additive constant. If this were not true, the interferometer would not work. All four red photons are said to "belong to the same state", which is a loose way of saying that they constitute a four photon system of entangled particles. *Entangled* means observations of them are highly correlated. But it does not explain why.

Mandel used a beam splitter to create two beams as did Michelson-Morley, but then split each beam with parametric down conversion which M&M did not do. This experiment has been done with Young's interference method which I will assume here because it is easier to diagram than the Michelson-Morley apparatus. The diagrams below are schematic: Do not use them in a technical manual on how to build interferometers. Viewed as electromagnetic waves, all the light rays, both UV and red, are coherent and phase locked producing interference fringes on the right and left screens. Young's interferometer automatically produces a band of interference fringes shown as a vertical band in the drawings. Mandel's interferometer would require some additional optics, but getting into such details would bog down the description so I am glossing over them.

Mandel's Experiment

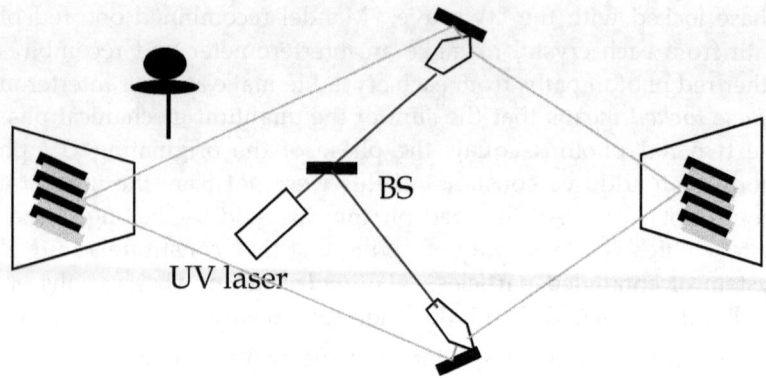

Light paths unobstructed: Interference at both screens

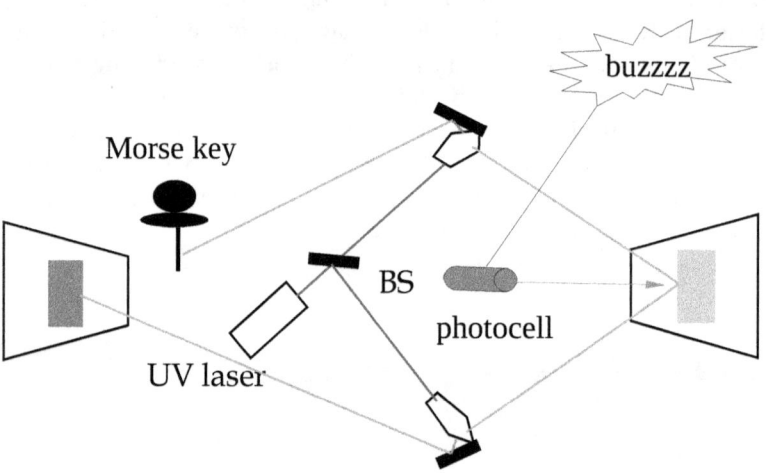

Any light path obstructed: Interference at neither screen
Less light at the obstructed screen

We make no essential distinction between the two red photon interferometers one on the left and one on the right. However, we follow the convention of labeling one pair of photons as the *signal* photons, arbitrary choosing the right screen. The left screen we label as receiving the *idler* photons.

A photon reaching the beam splitter BS has some amplitude to be transmitted by BS and some amplitude to be reflected. To keep things simple, we will assume they have the same magnitude. After passing BS, the photon is in a superposition of two states, both real (as in reality). The two subsequent paths are widely separated, say by about a meter in places. What we mean by *both real* is that if there is interference when the beams reach the screen we have to compute the amplitudes for a photon to get there taking **both** paths to each screen and add these amplitudes. Because the calculated amplitudes are signed numbers and we are assuming they are equal in magnitude, they can cancel. This happens in the center of dark fringes. For each path, the amplitude calculations involve exact descriptions of all configurations of optics along each path at the time the photon would have passed through – exact path lengths, exact window sizes, specifications of the precise optical properties and dimension of any glass the photon may have to pass through to get to the interference screen. This must be done for every path the photon can possibly take to get to each point on the screen. The amplitude for the photon to arrive at a point on the screen is the sum of all these amplitudes for all possible ways it could have gotten to that point.

Suppose we mess with the upper left red beam by poking something opaque into it. We will loose interference and one half the light. Light from the unblocked path is spread over the whole area without interference from the blocked path. **We would not expect that doing this would have any effect with the right interference pattern because we are doing nothing to its beams. Our physics teachers have been telling us: "If you don't mess with these beams you get interference fringes".** So Mandel said: OK, we won't touch the beams going to that screen but, anyway, we can find out which path the photon takes through that branch of the interferometer without getting anywhere near it, much less messing with its light paths. Remember what Feynman effectively told his students: *It appears that reality obeys one of two laws, but decides which one to obey depending on what you know.*

This is what Mandel, in effect, said: If the black thing is an efficient photon detector, and it stops a photon, we will know it came from the upper crystal, hence we have found out that the other red signal photon of the pair also from the upper crystal went to the right screen. This implies in the above diagram that the UV photon was transmitted by the beam-splitter and not reflected. If the photon detector registered nothing, but the CCD array on the left did register, we know its signal photon came from the lower crystal, so we know that the other red photon that went to the right screen also came from the lower crystal. This implies in the above diagram that the UV photon was reflected by the beam-splitter and not transmitted. <u>So we know which path every photon takes every time in **both** interferometers by just looking at what they are doing in the idler interferometer.</u>

Hence, we are able to know in which path the photon is going to the right screen without in any way messing with its photon paths of the right interferometer in the above diagram. Common sense tells us that we should still get interference at the signal screen because we are not in any way messing with the photons going there from ether the upper or lower crystals.

Mandel Found:

When <u>you</u> find out which path <u>any</u> photon took, both interference fringes at both the right and left screens disappear. In the above case, the right screen becomes uniformly illuminated – no interference fringes – but is twice as bright as the left because it is still receiving light from both crystals. We are not messing with either of its light paths. But somehow, although light is taking both paths just as it did before when we got interference, now the light waves do not interfere. But we did not do anything to these beams of the signal interferometer. We only looked at an idler beam. Half a century ago Prof. Feynman effectively told his students this:

Law 1. If **which path taken by the photon is known**, calculate the amplitude for the photon to take that path and square it, to get a probability the photon will arrive there and do this for every point on the screen. If it can take two different paths but "which path" it took is known, add the <u>probabilities</u> for both paths. **This law gives <u>no</u> interference pattern** because these probabilities are positive numbers and cannot cancel.

A physicist who repeated Mandel's experiment using Yong's inference, depicted here, described what he saw as "mind boggling". Elsewhere in this paper, the point is made that he was not being mind-boggled by physics, but by the collision between reality and his metaphysical beliefs.

By finding out what the idler photon did we know what its mate the signal photon is doing. Schrödinger coined the term "entangled" to describe this relationship which conjures up visions of some sort of spooky ectoplasmic fiber linking the two photons without explaining anything. Understanding what is happening is not hard, but this understanding comes at a price that many cannot bring themselves to pay: Stop believing the doctrine von Neumann described in this way: "Consciousness is in the physical world".

Mandel knew which path the photon took; what he <u>knew</u> destroyed the interference without any intrusion into either light path of the signal interferometer. How can just knowing something over here change how something happens somewhere else? Answering this question is the purpose of this epistle. Before discussing it, there is another consequence of Mandel's experiment that was understood before he did it, but had never been so spectacularly demonstrated.

Look Ma, no wires, no cables, no pull strings!

It was the demonstration of what Einstein with disdain called "spooky action at distance".

Just a piece of black paper

Alice's Morse Key

Now you can use this interferometer to make a two way Morse code link that sends messages across space with absolutely no physical connection and it is "instantaneous". The black flag on the Morse key is a piece of black paper in the upper left light path blocking the light path when you depress the key. At the right interference pattern, a photocell pointing at the center of a dark fringe is connected to a buzzer. Even though Alice used a black paper mask attached to the Morse key, she can, in principle, find out from which crystal the photon came. When she presses her Morse key, Bob's buzzer will sound, because the dark fringes at Bob's interferometer screen disappear. Will this really happen? Yes! This is what Mandel saw. We have non-local "causality".

Prof. Susskind's Lecture 6, *The Theoretical Minimum* of his series of online lectures on the quantum mechanics describes the simplest coupling between two systems, one of which we can think of as the "observed", and the other the "observer". He shows than even when these are no more than the simplest two state systems, some coupled states exhibit quantum entanglement. In this lecture he states that it is not possible to transmit information between remote entangled states and that such coupling does not constitute "non-locality". If these statements are understood to be true of any quantum mechanical system, they are false: It is certainly possible with a Mandel interferometer unless Mandel and his student's experiments were hoaxes. However, Prof. Susskind may be correct. Alice is not playing the game by his rules: She invented a rule Prof. Susskind didn't take into account: Smash the interferometer. That's why Bob stopped seeing interference fringes. But she didn't harm Bob's half of the apparatus: She only smashed her own. To understand what is going on quantum mechanically for Bob we need Alice's wave-function too.

It is very easy to fall into classical thinking trap when trying to understand what is going on, so easy in fact that Einstein did. It is not possible to transmit information at superluminal speed, as Einstein thought Copenhagen implied and to which he vehemently objected. In fact you could send it more quickly with an ordinary radio link. But you can send it with no link at all other than: The photons Bob and Alice observe are in states that had a common origin in the past in the form of single UV photons before they entered a beam splitter. Each photon exits the beam splitter in a linear superposition of two eigenstates which can become widely separated in space as in separate channels of the Mandel interferometer. It is meaningless to talk about what is "feels like" to be in each of these states: Although they all happen to the same conscious self (Tat tvam asi) they do not bind in awareness with the conscious experience important to the person reading this script: So we use the synonym of *consciousness* and say that each of these states of the photon is *real.* The existence of the photon in each channel also constitute awareness of which channel. The wave-function of the photon is now split into two spatially separated parts with non-zero amplitudes along the channels and that are extensive in time because the photons are highly monochromatic, or the interferometer would not work. If we integrate the Born probability density over phase space we will get 1. The space between the channels will contribute nothing: There is no amplitude for the photon to be there. Alice is on the threshold of opportunity to change the destiny of every one forever.

1. She can do nothing, in which case the wave function merges at the interference screens: In the dark fringes the amplitudes add to zero, and in the bright fringes they reinforce. It is meaningless to ask "which path did the photon take?" Look at the wave function: It took both.

2. She can poke something into one of her channels, say, just before the screen. What you believe will happen, depends on your metaphysical creed.

The photon wave-function must have a very small spread in wavelength, hence frequency, hence energy, for the interferometer to work. Its Fourier transform shows that the photon needs a lot of stretch room in time: Much longer than the time it takes light to travel the length of he apparatus. We cannot put obstructions in the time window without destroying interference. Believing that the photon really makes its journey at some precise moment but God, invoking the 11th Commandment of Uncertainty, prohibits us from knowing exactly when, is Copenhagen Physical World of Metaphysical Malarkey.

For the better part of a century, quantum math has been trying to teach us: There is no such place as the physical world; it is illusion created in the image of intuition hardwired into our central nervous system. Experience has never lied to us: It is never manifest except in consciousness. When Alice put a photocell into one branch of the Mandel interferometer, she destroyed both branches, because she then entered two states of consciousness caused by the eigenstates of the beam-splitter:

- the UV photon was transmitted and not reflected
- the UV photon was reflected and not transmuted

These two states do not bind in her awareness because they were caused by orthogonal eigenstates of the beam splitter. These were the states that recombined to cause interference because that combination was the wave function of both being transmitted and reflected. Now interference is impossible because Alice has destroyed the Mandel interferometer. Because states of consciously experienced awareness are real that know "which path" in <u>both</u> branches of what used to be an interferometer, but each state of awareness excludes the other. We have no need to call upon God to "collapse" one or the other of the above bulleted items because they are inconsistent now that we know which one the conscious self is in, tat tvam asi = that thou art, in both. They are only inconsistent in the Physical World. There is no such place. The states of consciousness of being aware of both are equally real: They do not bind in awareness because each was caused by orthogonal eigenstates of the beam-splitter. Interference is impossible in either because each excludes the other. Bob is the same conscious self as Alice. *Tat tvam asi*. Eventually, when the coherence time of each photon has passed, Bob will be in the same states of awareness as Alice caused by a mess of equipment made of lasers, beam splitters and mirrors that used to be an interferometer, now broken be Alice. Alice is <u>not</u> transmitting Morse to Bob with a piece of equipment invariant except by changes to its states. She is sending Morse to Bob by alternately wrecking and restoring an interferometer to send dot/dash on one hand and blank on the other.

Forget Morse: It's been decommissioned. Interrupt the beam with a chopper to send the individual bits of ASCII code so we can send text, sound and video images from one place to another with no wires, nor push-rods, nor pull chains, nor hydraulic lines. And no Goss ports. With absolutely **no physical connection at all**. And instantaneously. And in both directions. Is this all true? It is, if you stop and think what *instantaneously* means.

Slogging out of the metaphysical quagmire

This device is an interferometer. To be a good one we need the number of waves in each branch to be stable to about 1/10 of the wavelength which we take to be micron. If the interferometer legs are 1 meter, this amounts to 1 part in 10,000,000 = 1:1e+7. If they are L meters we need the wavelength to be certain to 1 part in L.10e+7. This means that the energy must be fixed to 1 part in L.10e+7 or the wavelength will be so incoherent the interferometer won't work.

The quantum mechanics Fourier transform complementarity relationship, between **energy** and **time** is $\Delta E . \Delta T \approx h$. If we know the time falls in some range ΔT then energy spreads out over a range ΔE. Energy and time are in this way are Bohr complements. Classical formulation, for classical conjugates, implicitly assumes $\Delta E . \Delta T = 0$. Knowing both with certainty results in as completely determined history of the evolution of the system. In the quantum mechanics the number of mathematically possible future histories is infinite. We are chopping a light beam to make binary 1s and 0s, or Morse dots and dashes, they will blur together unless ΔT is small enough, which makes ΔE larger.

Because:

$\Delta E . \Delta T / E = h / E$

$L.1e\text{-}7 . \Delta T = h / (h\nu)$

L.1e-7. $\Delta T = 1/\nu = \lambda/c$

$\Delta T = Le7.\lambda/c$

Using meters and assuming $\lambda = 1e{-}6$, $c = 3e{+}8$

$\Delta T = Le7.1e{-}6/3e{+}8 = L/3e{+}7$

This means that if L = 1 meter, if we chop the beam 30,000,000 times a second we will no longer have interference fringes – the interferometer won't work because the light is so incoherent in wavelength it cannot interfere. That's plenty of bandwidth to send high definition video. **And with no physical connection**. If L = 1 kilometer we are in trouble if we chop the beam 30,000 times a second. If we can get away with 10,000 we can't even send HiFi audio. But we could still send telephone quality audio 1 kilometer with **absolutely no physical connection between the transmitters and receivers**. To send information overseas, we are down to a few hertz: Now we can only send Morse code, but without any physical connection between the transmitter and receiver except: Two different photons had a common origin in the past.

This is where naive metaphysics traps even the greatest minds. Einstein didn't like the Copenhagen Interpretation because it sent information "instantaneously" in violation of the Special Theory of Relativity. But he didn't stop to ask what "instantaneously" means in the light of the quantum mechanics. If you are using a Mandel interferometer to send information overseas you can do it without any physical connection but you need an interferometer that stretches from one continent to another in which the path lengths remain stable to a ten millionth of a meter. And at that you can only turn the fringes on and off a few times a second and send Morse code. But "instantaneously". The length of a quantum mechanical "instant" is the length of time within which you cannot know when the photon made its journey without corrupting its wavelength so much the interferometer will not work. Roughly it is the time it takes light to travel the length of the interferometer, and to be safe, much longer. You can send video at the speed of light by radio or cable. Try sending audio with this Mandel telephone from Cornwall to Newfoundland and there are no fringes for a simple reason. A Mandel interferometer with a fast enough beam chopper to send audio overseas is **not** a Mandel Interferometer at all. No more is it a telephone. It is a Mess of Junk as Prof. Feynman may have called it: As a telephone it does not work and you cant fix it. If you could fix it, the quantum mechanics would be wrong. If you want to try fixing it anyway, try something easier first, like inventing perpetual motion.

The magic of quantum non-locality

As long as the data rate is slow enough to keep the Mandel apparatus operating as an interferometer, it is possible to send information from one point is space to another with absolutely no physical connection between them. **This is the magic of quantum non-locality.** Alice and Bob can talk in both directions with nothing connecting them but common events in their mutual past when UV photons passed a beam splitter. If data bits are passing between them, they cannot know when each bit was sent on a time scale comparable to the time light would take to pass between them. One is not sending information at superluminal speed. It could be sent more quickly, at the speed of light, using a modulated photon beam. One is lapsing into $h=0, c=\infty$ thinking if one believes: God knows exactly when each bit is sent but is keeping it secret. No secrets are being kept from us. Creating God's truth in the image of human belief is mankind's greatest failing.

We dice reality up into a coordinate grid in space with points in space and points in time. We justify this because we really can measure extremely small distances and very short time intervals and very exact values of energy. We then superpose this grid on a delusion: "physical reality" and are mind-boggled when it does not work.

Einstein did not take this into consideration. He fell back onto intuition, calling it "spooky action at a distance". He was taken in by the requirement that observing which path the photon took instantaneously destroyed interference implying faster than light causality. The paradox here is that Alice and Bob can chat with no causal link between them. This is inconsistent with the concept of physical reality. When Alice does something it "causes" Bob to experience something. And conversely. But with no causal link between them? Einstein, believing in causal locality, could not comprehend that such a thing was possible.

Causal inconsistency

It was known from the earliest days of the Special Theory of Relativity that superluminal information transfer can cause inconsistency in causality. You can cause your parents not to conceive you in which case you won't be around to cause them not to. You could do this in principle if there were an intergalactic superluminal telephone system where you could route calls through distant galaxies and talk to your parents before you were born. The galaxies would have to have the correct relative velocities to make the call arrive at your parents say a few months before you were conceived. A theorem proved by Eberhard and

Ross shows that if the equations of relativistic quantum field theory are correct, it is not possible to violate causality using quantum effects: Eberhard, Phillippe H.; Ronald R. Ross (1989). "Quantum field theory cannot provide faster-than-light communication, *Foundations of Physics Letters* 2 (2): p. 127–149.

Knowing "which path" destroys interference

Following the above diversion into quantum non-locality, we now return to the central question: How can we understand how knowing which path the photon took destroys interference? This requires understanding the gist of Everett's theory. Everett did not commit to a definite interpretation of reality, leaving the question open. He specifically left it open to consciousness by stating that his formulation did no violence to the doctrine of psycho-physical parallelism. In the following description of the core of his conclusion I do commit to the interpretation of reality as consciousness; that which is knowable, such as recorded data, is awareness. In principle, any reference to consciousness can be omitted, using Everett's mathematics as a descriptor of awareness alone. The reason consciousness need not be considered is that it does not exist. Although it is incontrovertibly real to you, it is not scientifically demonstrable. However, if consciousness is not held in mind, the reader is apt to interpret awareness as describing some preconception of reality. This invariably will be the "physical world" of intuition which, successfully, does not admit inconsistency in awareness. However, it leaves leaving one awash in paradox that can only be removed by adding a metaphysical patchwork: Collapsing the unwanted inconsistencies and abandoning causality.

Knowing "which path" does not necessarily change the photon. It changes **you**. Von Neumann's accounting of this change was approximate and not correct. It was correctly described mathematically by Everett. Mandel's experiments and the *Double-slit quantum eraser* experiment described below show that the photon (in these cases) was not affected, but the experimental results were. **We have to learn to recognize that we are quantum mechanical systems and experiments designed not change the system we are observing, even if it is a simple photon or electron, can radically change, forever, what happens to us.**

Seeing interference in Mandel's interferometer

If we want to see any interference at any screen we have to configure the apparatus so that there can be no causal consequence by the eigenstate in which the photon is transmitted and not reflected nor by the eigenstate in which the photon is reflected and not transmitted within the coherence time of the photon. The reason for this is that awareness results from the photon not entering one state and not entering the other when interference is a causal consequence of both. But interference requires that awareness is caused by both, following the simple law that the amplitudes by each be added. If one or the other amplitude is zero, there will be no interference. The quantum mechanics tells us in simple terms what the ultimate states of consciousness will be after the dust settles well outside the coherence time and space of the photon's existence. You may be very curious what the wheels and gears are doing before the dust settles. This question cannot be answered without poking probes into the apparatus, but that is a different experiment. It is easy to imagine the impossible. For example, if two electrons a and b collide head on and two electrons c and d emerge from the collision, we can pretend we can watch the collision like marbles and see that a is d and b is c. But this is impossible. a and b are indistinguishable as are c and d. It is a mistake to try to make a physical model, only to be confounded by paradox. Here, Feynman's offers good advice: do not ask 'But how can it be like that?' if you are trying to look inside Bob's boxes to understand what is happening in them before the dust settles. That is what we do when we poke stuff into the particle's path to see which path it follows and when it was here or there. We may not doubt that Newton wondered 'But how can it be like that?' when he looked at his simple second and third laws of motion and his simple law of gravitation and knew they predicted the motion of the planets to all observable accuracy**. He may have asked himself: What wheels and gears does God use to make these laws work? Now we can answer this question: His laws were mathematical approximations to deeper truths that we came to understand in the first third of the 20th century: The quantum mechanics and both theories of relativity. Newton could never have inferred them by wondering: 'But how can it be like that?' We only came to understand them, and that was not easy, when Newtonian mechanics, classical electromagnetics and kinematics made predictions that were demonstrably wrong and supplied the clues of what went wrong.

**Footnote: Newton was not certain that his laws were correct and wondered if they only worked correctly in the simple case of two bodies like Mars and the sun. He tested them on the sun, earth, moon system discovering anomalies in the moon's motion unknown to the ancient astronomers who knew only of those that revealed themselves through the timing of eclipses. However his anomalies were wrong by small factors like 2 or π. He never knew whether this was a sign that his mechanics was in error. Today we know that these factors were due to approximations he had to make to render the problem tractable.

That he was able to do as much is a monument to his genius: The earth-moon problem was not solved to observable accuracy until the 20th century. Newtonian mechanics predicts the motion of all the planets and their moons to present day observable accuracy except for a small anomaly in the orbit of Mercury that Einstein explained.

Understanding what Alice and Bob see

Mandel's experiment is the watershed of belief in what is real. We start by rejecting Feynman's bluff: ***Nature sees no paradox.*** <u>Nature</u> never does, Richard. ***But it is very peculiar.*** It boggles your mind, doesn't it? Which means in plain language that we <u>do</u> see it as a paradox.

Saying that Bob and Alice are the same conscious self, ***tat tvam asi***, says nothing science can grasp because conscious selfhood, although we all experience it, cannot be observed. We have to come up with an Alice and Bob who meet the ***tat tvam asi*** requirement and be the same self in awareness that will meet Prof. Gazzaniga's conditional approval. She is SuperAlice who can morph into two identical looking twins, Alicia and Bobbie. Alicia has a left brain. Her right hemisphere is replaced a two way radio transceiver connected to the corpus callosum. It is in constant contact at the speed of light with a similar radio in Bobbie's left chamber connected to her right brain brain through the corpus callosum. They will have trouble walking, so they will have to stay in wheel chairs for the duration of the Mandel experiment. Unused nerve endings like the optic nerves to the radio hemispheres will be capped off (until she morphs back to plain old Alice). Alicia watches the Alice interference screen in her apparent right visual field of both eyes, and Bobbie watches the Bob screen in her apparent left visual field. Alicia has her right hand on Alice's Morse key.

The transit time of the SuperAlice corpus callosum is shorter than the lifespan of each photon wave function necessary to have monochrome photons necessary for interference. Alicia can in principle find exactly when the photon hit the blocking device, and call that time T. It is a spike in amplitude. OK, take the Fourier transform of that spike and the photon energy and hence wavelength is spread over hell's half acre and no interference is possible. This is what the mathematics of reality has taught us and it is not physical world math.

When SuperAlice comics come out, you will read how the mad scientist Gazzaniga created her. He has a remote controller in his pocket and can turn off her corpus callosum link at will. When he does it has no effect at all on the conscious experience of SuperAlice. Alicia sees no interference fringes at her screen and Bobbie none at her screen either. With a name more terrible than Frankenstein or Dracula, Gazzaniga is yet a kindly guy (his picture is at the end of this book) who will leave a girl in a state of bewilderment only just so long wondering why her interference fringes suddenly disappeared. Reaching in his pocket he will turn on her corpus callosum back on. Then it will be obvious that her right hand, Alicia's right hand that is, is holding a piece of black paper in the interferometer light path, putting SuperAlice into an eigenstate of the beam splitter. SuperAlice's escapades in the other eigenstate are documented in DeWitt/Graham Gang Other World Adventure comics which, unfortunately, are not obtainable in this universe.

It is a mistake to try to imagine as we would with a physical world model what is happening at each instant of time. The life of the photon takes up a lot of time room. We can not ask without absurdity where Mt. Everest is located to the nearest millionth of an inch. Mt. Everest takes up a lot of space room. There is no experimental way of asking what is happening instant by instant without changing the experiment in a way that would destroy interference. Alice did destroy interference by asking to be aware which path the photon took. Everett tells her that the answer is the same as before: The photon takes both paths it took before, but now Alice, *tat tvam asi*, enters two states of consciousness of being aware of these two paths. But they do not bind in awareness. In each path the amplitude to be in that path is 1 and the amplitude to be in the other is 0. So there can be no interference.

Alice and her photocell thus have veto power over interference. She knows how to switch between Feynman's third and second laws described earlier: By knowing what Mandel's student Banning called "which path". Thus Alice can send signals to Bob even though there is no physical connection between them. If Einstein saw this it would be interesting to ask him if he thought it was spooky action at a distance. If Alice and Bob's communication were possible because they were entangled with ectoplasmic fibers that passed through solid walls so Alice could signal Bob by tugging strands of superluminal ectoplasm, it would be spooky. But I think Einstein would be happy to know that information cannot be transmitted at superluminal speed and happy to leave ectoplasm where it belongs: In spiritualist séances. There is no mystery to what is happening when we understand that when Alice blocks a beam-path the conscious self enters two states of consciousness that do not bind in awareness. 1 The photon was transmitted by the beam splitter and not reflected. 2. The photon was reflected by the beam-splitter and not transmitted. No philosophy which respects the ultimate reality of the physical world can accept this, because it makes this equation: reality = existence.

Banning

Talking about a single photon and consciousness may seem absurd. Consciously experienced awareness is associated with complex brain function. The only important distinction to be made between our brain of zillions of electrons, quarks and photons on one hand and one photon on the other is that the brain can be aware that it is conscious, the photon is not and not even aware of its own existence.

Interesting experiments were done by David Banning, a student of Mandel and described in Banning's PhD 1998 thesis at the University of Rochester, New York.

What destroys the interference at the screen that had no obstruction in its light paths? Here is an answer we got from Banning and one we may have gotten from you:

1. <u>Interference is destroyed by the light entering into two eigenstates each of which can have different causal consequence after leaving the beam splitter.</u>

2. You may say: "That's absurd. Unless the light has a finite amplitude to take two or more paths you don't have an interferometer. And hence no interference is possible. You have to have some sort of beam splitter if you want to see interference fringes. What do you mean beam splitters destroy interference?" you may well ask.

Both of these two statements are true. They are only apparently contradictory.

Banning's point is that as long as one is aware the photon is in two orthogonal states that embody "which path" information, interference is impossible. We can only regain interference by destroying the "which path" information. He describes a number of experiments in which this information is destroyed and interference restored.

For present purposes, instead of adopting Banning's historical "which path" jargon, I would rather describe the two eigenstates of the photon as

1. transmitted by the beam splitter and not reflected
2. reflected by the beam splitter and not reflected

Both of these states being equally real. The photon in each of these states constitutes states of awareness that are mutually exclusive. We are talking about consciousness and awareness involving a single photon, not an entire human brain. While a single photon is not a human brain, each of these states could, but need not, cause real states of human consciousness in the human brain that constitute "which path" knowledge, but these states of consciousness do not bind in awareness. We need not concern ourselves with how it feels to be a photon.

<u>Doing so does not "destroy" interference.</u> When you put the photocell in a light path, **you asked** to be in a state of awareness ("world") in which there **is no interference**. So, if one asked to be in either of the above states of awareness by putting photocell in the light path, why does it boggle ones mind to find no interference? If you catch a train to Pretoria, why would it boggle your mind when you get there to find that you are **not** in Trenton, New Jersey?

The paradoxes of Mandel's interferometer arise from trying to understand what is happening by seeing the photons as elements of a physical world.

Electron Spin

Appendix F describes eigenvectors which are referred to here.

To illustrate how quantum math works and how even the simplest conclusions are colored by metaphysical belief, we take the case of electron spin. This case is a good illustration of the mathematics because it can be understood even if you only understand the elementary grade school arithmetic operations and no algebra at all and forgot your multiplication tables. (If you don't know what they are, ask your granddad) Although it took the genius of Nobel Prize winners to discover <u>how</u> nature works in this case, what nature does is simple. If you want to know <u>why</u> nature works this way, you will have to ask God, or Mother Nature. If God does not answer your prayer, I can't help you. If you ask Mother Nature, you are in for a rough ride. Planck, Einstein, de Broglie, Bohr, Heisenberg, Born, Pauli, Schrödinger and Dirac did ask mummy and it took them three decades to understand her answer. They argued and bickered about reality until Everett came along. He looked at her math and realized it revealed something amazing.

Stern-Gerlach Experiment

To set the stage, we relate the story of a landmark experiment of quantum history. Unlike the polarization of light experiment which involves similar math, this experiment is not one you can do at home. Stern and Gerlach set out to measure the magnetic moment of silver atoms. Had the experiment gone as expected, the results, of technical interest only to small group of physicists, would have been buried in the dusty archives of physics and forgotten after more accurate measurements of the magnetic moment of silver were made. The experiment was a success, but things did not go as expected. Its greatest success was something Stern and Gerlach never dreamed would happen. Silver atoms were shot out of a gun in a vacuum in the form of a narrow beam that struck a glass slide, where they stuck to the glass. The narrow beam of atoms passed over and along a knife edge magnet that made a magnetic field that was highly non-uniform in the vertical direction. This would attract the atom down if its the silver atom magnetic pole pointed down or repel it upward if it pointed up. Before getting shot out of the gun, the silver atoms, boiled off as a vapor, had random orientations, so atoms oriented sideways in the magnetic field would experience a small twist but there would be less or no net upward or downward force: Or so reasoned Stern and Gerlach, guided by classical physics. Because all orientations of the atom were possible, they expected to see a vertical streak of silver on the glass plate after the running the experiment for some hours. The length of the streak would tell them the magnetic moment of silver. But there was no streak. The story goes that Stern smoked stogies and breathed cigar smoke on the slide while looking for the silver streak he couldn't find. The silver was there but not where he expected to see it. The stogie smoke turned the silver black – then he saw two black silver marks, one at the top and one at the bottom of where the silver streak was expected, and nothing in between. Stern was looking at the two eigenstates that answer the question "Is your orientation vertical?". The answer is always yes: It is either "up" or "down" and nothing else. Had the magnet been horizontally oriented, the question would have been "Is your orientation horizontal?". The answer is always yes: It is either "left" or "right" and nothing else.

Imagine the silver atom Ag teaching Stern physics. The silver atom is giving an honest answer to every question; it has not been hornswaggled by Copenhagen.

Stern: Is your spin aligned with my magnet?

Ag: Yes

Stern: Up or down?

Atom Yes

Stern. It cannot be both. Which?

Ag: Both are real.

Stern: I only saw down.

Ag. I told you so.

Stern: I am turning my magnet sideways. What are you going to say now, eh?

Ag: Yes

Stern: Left or right?

Ag: Yes.

Stern. It cannot be both. Which?

Ag: Both are real.

Stern: I only saw right.

Ag: I told you so. And stop blowing stogie smoke at me, I just got polished.

Classical physics, intuition and common sense all find this result incomprehensible. We have to face the fact that this is how reality works.

The magnetic moment of a silver atom is locked to its spin or intrinsic angular momentum which we see as a vector in three dimensions. Electrons would give similar results. We think of their spin angular momentum which is always h/2 as being around an axis which points somewhere in 3D space. However, we cannot experimentally ask the question "which direction are you pointing?". We can only ask "are you pointing along this line". The answer is always yes. When we try to understand why, we risk stepping into metaphysical no-mans-land. Suppose we call the two directions along the line *this* and *that*, avoiding *up* and *down* or *left* and *right* to avoid being prejudiced. Suppose when we observe an atom it says *that*. If we ask the same unchanged atom the same question again it will always say *that*, as though the question is redundant because we already know the answer.

The Stern-Gerlach experiment cannot be done with electrons because they are charged and are strongly deflected passing through the magnetic field. In fact, Bohr and Pauli believed that measuring the magnetic moment of an electron by any method similar to Stern-Gerlach was impossible. Revered as the opinions of these savants were, they cast an evil spell, as did Newton when he proclaimed that light did not propagate as a wave. In recent times H. Batelaan, T. J. Gay, and J. J. Schwendiman of the University of Nebraska, Lincoln expressed the view that a tabletop Stern-Gerlach electron spin filter is feasible, albeit one using a different magnetic field arrangement where the electron beam travels not across, but along field lines that are also very straight. The technical difficulties of doing such an experiment won't bother us here. We will assume we can sort beams of electrons into two beams for the two spin states of each 3D direction. We will consider three orientations of the Stern/Gerlach magnets: In the Z, X and Y directions

The vectors we talk about only have two numbers and the matrix operators composed of only two vectors. We take three orthogonal directions in ordinary space and assume that we have Stern-Gerlach type devices to measure the spin of an electron in any of these three directions.

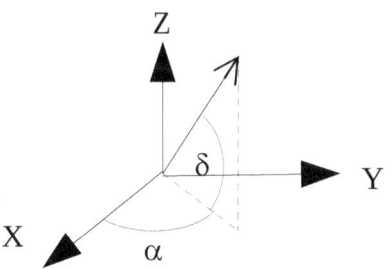

We have only shown the positive directions to simplify the diagram. This is called a right handed system. We could use the mirror image just as well which is left handed, but agree not to. Note that pointing directions in three dimensions is only a two dimensional problem. An astronomical telescope or surveyors transit can point in any direction in three dimensional space by specifying only **two** angles shown as α and δ in the above diagram. The Pauli spin matrices have only two degrees of freedom, but there are three of them not counting the unit matrix.

Pauli spin matrices

These Hermitian matrices are quite simple and present a good way of getting the flavor of matrix mechanics. Good discussions of how they are used is given in Feynman's *Lectures on Physics Vol III* and on the web in YouTube videos by Bob Eagle a the website DrPhysicsA. Enter *DrPhysicsA Electron Spin* on your browser. However, such traditional discussions of these or any other mathematics of the quantum mechanics will drag in metaphysics presented as physics, which is the very metaphysics this screed it trying to abolish. For this reason I deferred this discussion to after the chapter on consciousness.

We hand Wolfgang Pauli a Nobel Prize and he hands us back these three 2x2 matrices:

```
         σz              σx              σy
       |1   0|         |0   1|         |0   i|
       |0  -1|         |1   0|         |-i  0|
```

Each of which has two eigenvectors referred to the z states +z as you can see from the Z case. (Please read these as column vectors).

```
   Zu      Zd         Xu         Xd            Yu          Yd
  |1|     |0|      |1/√2|    | 1/√2|        |1/√2|     | 1/√2|
  |0|     |1|      |1/√2|    |-1/√2|        |i/√2|     |-i/√2|
```

Each of which has these eigenvalues for the columns above.

```
   1      -1           1         -1             1          -1
```

Nomenclature: It is conventional to write vectors in matrices in rows and the vectors they multiply in columns, as above, to simplify the algorithm to do multiplication. Writing vectors in rows or columns does not change their mathematical meaning. We use rows below to save space.

If you multiply any of these eigenvectors by its own matrix, you will get the same eigenvector back multiplied by its corresponding eigenvalue. This is the property of Hermitian matrices which is what these are: If you transpose such a matrix (flip it about its upper left to lower right diagonal) and take the complex conjugate, you get the same matrix back. To get the complex conjugate of whatever complex mess change the sign of i wherever, if ever, it appears. That's the complex conjugate of whatever.

The above eigenvectors refer to the z spin components as the base states in all cases. This choice is arbitrary. If you wanted to refer to the x states, the matrices do not change. Only the names do. Nature does not change her rules when you rename things.

Caution: In some books you will see the above eigenvectors referred to as Z+, Z−, +z and so on. This terminology is dangerous. We have called them Zu and Zd etc as Stern and Gerlach may have called them the *up* and *down* states. If you just look at Zu and Zd and take their scalar product, you can see that is is 0 because these two vectors are orthogonal to one another. So are Xu .Xd = 0 and Yu . Yd. = 0. To be quantum mechanically copacetic, you should evaluate Dirac's

<bra|ket> = 0

which is exactly the same thing, except the complex conjugate is used in the `<bra|` case which, in the three cases at hand, changes nothing.

Sloppy notation could lead one to conclude that Zd written as Z− is the negative of Zu written as Z+. Furthermore, the scalar products of unit vectors and their negatives is -1, not 0.

We will consider three orientations of the magnet of the apparatus along the three orthogonal directions of 3D space of the above diagram Z, X and Y. The Z magnet will split the beam into two beams called u and d. An X magnet will deflect it left and right but we still call it u and d and the Y magnet will deflect it in and out and we still call it u and d because these are directions relative to the magnet assembly regardless of how the apparatus is oriented in space. To do the Y experiment we would have to change the beam direction by a right angle without changing the electron spin. We won't worry about how that is done. Feynman says it's OK to do this if your are wearing a mathematician's hat: It absolves you from worry about how to build the apparatus. But please: keep your math hat <u>on</u> to protect you from being pelted by a barrage of test-tubes and Bunsen burners by the lab crew.

It may appear that the Z states are orthogonal to the X and Y and X and Y to one another, because the magnets are spatially orthogonal in the Z, X and Y positions. This is not true: Any *u* or *d* state resolves with finite amplitude onto any other *u* or *d* state. Orthogonality is only true between a *u* and *d* state of the same magnet orientation. This is an example of the marked differences between Hilbert Space and ordinary space we spoke about earlier.

If you want to find the amplitude that a system in **this** state will be measured to be in **that** compute the Dirac <bra|ket>, the inner product of bra and ket. This tells us how much the |ket> projects onto the <bra| direction. (We have to take the complex conjugate of the vector used as bra. This is done so as to make the projection a real number = 1 if we project a unit vector onto itself).

<that|this> = (that*).(this) = scalar product of (that*) and (this)

that* is the complex conjugate of **that** and the . means inner product. Remember that (a,b).(x,y) = a.x + b.y is the definition of scalar product. It is the projection of (a,b) onto (x,y) or conversely. We are using Dirac's <bra|ket> notation. Say "Thank you uncle Paul", and give him a Nobel Prize. No, take is back – he already has one – he invented this notation later.

If you do this for any eigenstate and itself, you will get an amplitude of ±1 which squares to **1**. This is the Born probability of certainty that a system in an eigenstate will be measured in the same eigenstate. This is called *normal*. If you do this where **this** and **that** are two different eigenstates, <that|this> will give you 0, which means that these eigenstates are orthogonal. Since the eigenvectors are *normal* and *orthogonal* we call them *orthonormal*. Dirac used fewer words: If $e_1, e_2, e_3, \ldots$ are eigenvectors:

<e$_i$|e$_j$> = $\delta_{i,j}$ = 1 if i=j and $\delta_{i,j}$ = 0 if i≠j

$\delta_{i,j}$ is called the Kronecker delta.

Even if you flunked grade school arithmetic, you now know enough Nobel Prize winning quantum mechanics mathematics to separate reality from metaphysics. If you are up to your knees in quag, you have to climb out first: The first step is the realization that when we talk about "up" and "down" in the Stern/Gerlach experiment, these are where the silver atoms went relative to the magnet, we are not talking about the two different directions of a geometric vector, but two different eigenstates of the electron spin: They are orthogonal to one another like all eigenstates of a particular measurement are to one another.

Suppose you have a beam of electrons and know that the spin vector of all of them is pointing in the Zu direction, say because you prepared them in that direction with a magnetic field oriented in that direction. We now take this upper beam and analyze it with horizontal Stern-Gerlach magnets that ask the question: is you spin vector pointing is the

+X or -X direction. Intuition tells us the answer is neither. Its pointing +Z and not at all in the X or Y direction. Hey, if a helicopter takes off and flies straight up and we ask the pilot "how far north or south have you flown?", the answer is neither. If we ask "Well then, how far east or west have you flown?" the answer is still: "Neither. I am going up".

This is not true of electron spin. We can calculate the amplitude for the electron spin in the Zu state vector (+1 0) = |ket>, to be measured in the Xu state vector (1/√2 1/√2) = <bra|

$$<Xu \mid Zu>$$
$$<(1/\sqrt{2} \quad 1/\sqrt{2}) \mid (+1 \quad 0)> = 1/\sqrt{2}$$

This is the amplitude. We square it to get the Born probability, giving ½. In this book, the word *probability* is sacrilege, so we call it the Born weight.

We also calculate the amplitude for the electron spin to be in the Xd from the Zu state vector (+1 0) = |ket> and the Xd state vector (1/√2 -1/√2) = <bra|.

$$<Xd \mid Zu>$$
$$<(1/\sqrt{2} \quad -1/\sqrt{2}) \mid (+1 \quad 0)> = 1/\sqrt{2}$$

This is the amplitude. We square it to get the Born ~~probability~~ weight giving ½. Again. The sum of these ~~probabilities~~ weights is 1 as we expect and they happen to be equal.

We now put Y analyzers in both output ports of the X analyzer:

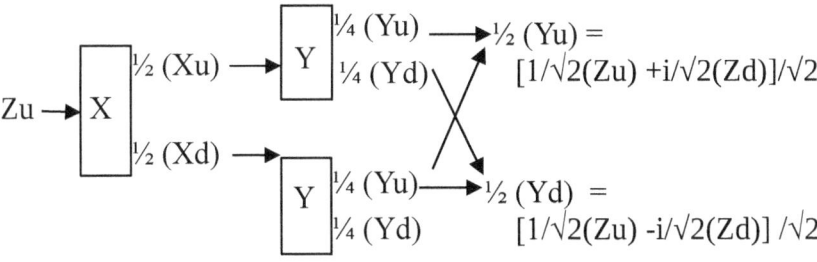

so we end up with four beams with the electron beams in the Yu and Yd states. The fractions before each state in the above diagram left of the () are the Born ~~probabilities~~ weights to be there.

211

Here is the math:

$\langle Xu | Zu \rangle$

$\langle 1/\sqrt{2},\ 1/\sqrt{2} | 1,\ 0 \rangle = 1/\sqrt{2}$

This is the amplitude to be in the Xu state which is thus

$1/\sqrt{2} | 1/\sqrt{2},\ 1/\sqrt{2} \rangle = | 1/2,\ 1/2 \rangle$

$\langle Xd | Zu \rangle$

$\langle 1/\sqrt{2},\ -1/\sqrt{2} | 1,\ 0 \rangle = 1/\sqrt{2}$

This is the amplitude to be in the Xd state which is thus

$1/\sqrt{2} | 1/\sqrt{2},\ -1/\sqrt{2} \rangle = | 1/2,\ -1/2 \rangle$

Now each of the above two states enters a Y analyzer with these four results:

For the first Y analyzer:

$\langle Yu | 1/\sqrt{2}\ Xu \rangle$

$\langle 1/\sqrt{2},\ i/\sqrt{2} | 1/2,\ 1/2 \rangle = (\sqrt{2}/4 - i\sqrt{2}/4)$

This is the amplitude to be in the Yu state from Xu which is thus

$(\sqrt{2}/4 - i\sqrt{2}/4) | 1/\sqrt{2},\ i/\sqrt{2} \rangle =$
$| (1/4 - i/4),\ (1/4 + i/4) \rangle$

The quantities in () are not vectors but complex numbers.

$\langle Yd | 1/\sqrt{2}\ Xu \rangle$

$\langle 1/\sqrt{2},\ -i/\sqrt{2} | 1/2,\ 1/2 \rangle = (\sqrt{2}/4 + i\sqrt{2}/4)$

This is the amplitude to be in the Yd state from Xu which is thus

$(\sqrt{2}/4 + i\sqrt{2}/4) | 1/\sqrt{2},\ -i/\sqrt{2} \rangle =$
$| (1/4 + i/4),\ (1/4 - i/4) \rangle$

We now do the math for the second Y analyzer:

$\langle Yu | 1/\sqrt{2}\ Xd \rangle$

$\langle 1/\sqrt{2},\ i/\sqrt{2} | 1/2,\ -1/2 \rangle\rangle = (\sqrt{2}/4 + i\sqrt{2}/4)$

This is the amplitude to be in the Yu state from Xu which is thus

$(\sqrt{2}/4 - i\sqrt{2}/4) | 1/\sqrt{2},\ i/\sqrt{2} \rangle =$
$| (1/4 + i/4),\ (-1/4 + i/4) \rangle$

Remember: Quantities in () are not vectors but complex numbers.

$\langle Yd | 1/\sqrt{2}\ Xd \rangle$

$\langle 1/\sqrt{2},\ -i/\sqrt{2} | 1/2,\ -1/2 \rangle = (\sqrt{2}/4 - i\sqrt{2}/4)$

This is the amplitude to be in the Yd state from Xu which is thus

$(\sqrt{2}/4 + i\sqrt{2}/4) | 1/\sqrt{2},\ -i/\sqrt{2} \rangle =$
$| (1/4 - i/4),\ (-1/4 - i/4) \rangle$

Now we will combine all the beams coming out of the two Y analyzers into one beam. Don't worry about how to do that. Making sure our math hat is on, just tell the lab crew to do it. Remembering that these vectors refer to the Z basis. Suppose we put a Z analyzer at the output of the merged beams, what will we see? We will do the math; Here it is:

```
|(1/4-i/4),(+1/4+i/4)> +
|(1/4+i/4),(+1/4-i/4)> +
|(1/4+i/4),(-1/4+i/4)> +
|(1/4-i/4),(-1/4-i/4)> =
-----------------------------
|1, 0>    But we started with this Zu state!
```

After all that mishegoss, the electron has never forgotten its original state. But when it gets back into its original state, it has forgotten all the mishegoss it went through. This would be true no matter what its original state was.

The above is physics. Now, lets talk metaphysics. The Copenhagen Interpretation has no choice but to recognize <u>for each electron</u> that the electron in reality takes all of the above paths. We could make life difficult for the Copenhagen or any other physical world philosopher by asking "How do you reconcile this with the concept of physical reality?" Instead, make life even more difficult for them by doing this: Make some very tiny coils of very fine insulated wire like this:

Bend these into circles which I can't do with this drawing tool. This makes a coil torus which is the technical term for a doughnut. This is called a toroidal inductor. I will just call it a toroid.

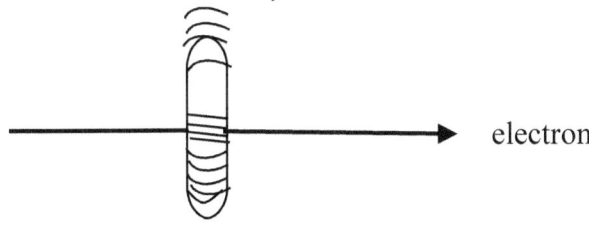

OK, OK. Next time you wind the coil.

An electron going down the axis of the toroid will cause a small pulse of voltage at the torrid inductor terminals which are connected to a high input impedance amplifier. I am not an instrumental design engineer, but will assume it can be engineered so that the passage of an electron through the torrid will not alter the spin state of the electron. The toroid does not detect the electron spin. It only detects the passage of the electron's charge which amounts to a small pulse of electric current through the toroid, causing a small induced potential pulse between the coil terminals. This would not work with photons (or silver atoms) which have no net electric charge. We could have the amplified voltage pulses store data in computer memory identifying what torrid sent the pulse and when it was sent, and then never look at the data, so nothing would alter human consciousness. Doing this would not alter the results of the experiment. However, I will assume that someone does become humanly aware of each torrid signal as soon as it happens. To ensure that this awareness enters human consciousness, I will assume that the observer is you.

We put one of these toroidal detectors in one of the Y analyzer exit paths and send another electron. With each such electron we will be aware of one of two possibilities:

1. The electron passed trough the torrid and exited in the merged beam

2. It did not, but exited in the merged beam anyway

We are about to see why the quantum mechanics is a theory of awareness as Everett forecast. It is not directly a theory of consciousness. We need consciousness as reality only to prevent meta-physicists from injecting an implied pseudo reality into their interpretation. This messes up the physics by confusing awareness with reality: For example, introducing randomness and superluminal eigenstate collapse into a completely deterministic theory. I cannot prove to you scientifically that consciousness is real, but I can describe in a single word an experiment that does: You.

This experiment is a parallel to one described later in which Leonard Mandel could tell which path a photon took through an interferometer, not by looking at what happened inside it, but by looking at something else happening external to it. Here we are not looking at electron spin inside the apparatus, but by looking at the eigenstates determining where the electron charge goes or does not go. Doing so changes the spin we observe, by looking at something else that does not change the spin. Understanding what we see will allow us to dredge a lot of metaphysical muck out of Feynman's drain. It was not put there by Feynman himself, but it is muck he gave those who, having the courage to ask: "But how can it be like that?" went down that drain.

We will see below that what is happening is staring us in the face from the page of the above math, but we are blinded by metaphysical prejudice. Everett was first to see it in pointing out that in reality the electron did follow all paths. If we explicitly equate consciousness with reality which Everett did not do, and recognize that all these states are simultaneously real. They do not bind in awareness because each states is caused by a different eigenstate of the electron and the eigenstates are mutually orthogonal. Difficulty in understanding this is caused by belief in the ultimate reality of the physical world of classical physics and intuition. This belief must be laid aside. The toroids had no effect on the electron spin. Yet when we merge all the beams, we will not find the electron in a perfect Zu state. How can this be if the toroids didn't change the electron we are observing? Here is the reason: Please do these two sums supposing that the torrid was placed in the third channel:

Case A. The electron was seen in channel 3 and hence could not have been in any other, so we gray out those cases:

~~| (1/4-i/4) , (+1/4+i/4)> +~~
~~| (1/4+i/4) , (+1/4-i/4)> +~~
| (1/4+i/4) , (-1/4+i/4)> +
~~| (1/4-i/4) , (-1/4-i/4)> =~~

| (1/4+i/4) , (-1/4+i/4)>

The Born ~~probability~~ weight to be seen exiting in a Zu state is 1/8 and the weight to be in a Zd state is 1/8. Reason: Evaluate <bra|ket| with bra = Zu = <1,0| and |ket> = the above vector. The fact that Zu = |1,0> and Zd = |0,1> makes it easy: Just square the coefficients of 1 and i and add them to get the ~~probability~~ weight.

Case B. The electron was not seen in channel 3 and hence could not have used that channel but did use all the others, so we gray out case 3:

| (1/4-i/4) , (+1/4+i/4)> +
| (1/4+i/4) , (+1/4-i/4)> +
~~| (1/4+i/4) , (-1/4+i/4)> +~~
| (1/4-i/4) , (-1/4-i/4)> =

| (3/4-i/4) , (+1/4-i/4)>

The Born ~~probability~~ weight to be seen exiting in a Zu state is $5/8$ and the weight to be in a Zd state is $1/8$.

When we add all these weights we get 1: The electron did not get lost in the apparatus, and had to come out in some state or another. The total weight to exit in a Zu state is ¾ and a Zd state is ¼ for both cases A and B, if we repeat the experiment many times. ¼ of the time we will detect the electron passing the toroid in channel 3.

If you use Everett's interpretation, every time you send an electron through the apparatus, exactly the same thing happens. The conscious experience of Zu and that of Zd are both real, but they do not bind in a common state of awareness. If you repeat the experiment many times and you are aware of Zu last time, it can be Zd this time. These are different pathways in awareness through consciousness, which does not exist, but is real, and not things happening to The Physical World which does exist in the world of physics before Michelson and Morley and the Ultraviolet Catastrophe, and which is not real.

Observing the electron did not change the electron spin. We only observed which channel it passed through, or did not pass through, by looking at its charge. But doing so did change to spin state of the electron at exit from the apparatus. The explanation is simple: Looking at which channel it took or did not take did not change the electron, but it did change **YOU**. This cannot be understood in a physical world model. You changed your journey through consciousness. Suppose you did not turn on your toroid signal amplifier so you, Sally Jones, could not be aware which way the electron went, would all the electrons have reverted to all exiting in the Zu state again? If Feynman's rules are correct, the answer would be that it would revert to the Zu state if, after the dust settles, you cannot, even in principle, determine whether the electron went through the torrid or not. The way to answer this question is to do what Prof. Feynman did to determine which path a particle took through an interferometer. The more successful you are at doing that the more you destroy interference. Here, electrons take multiple paths through the apparatus and "interfere" to make a Zu state at the exit. The more successful we are at finding out what the electrons are doing inside the apparatus the more we mess this up.

Although the interaction between the observer and the observed system was considered by von Neumann, the compete analysis of the interaction between them is due to Everett.

Arrow of Time

It is blatantly obvious to every three year old that there is asymmetry in time. She can remember yesterday but she can't remember tomorrow. When she is four she notices dad complaining while reading the stock market report in the evening paper that he is sorry he did not get this paper yesterday, but he never gets tomorrow's paper today. The family photo album only shows pictures of the past, never the future.

When she asks her parents why, she is told that it is because the *arrow of time* points from the past to the future. You can remember yesterday because it happened already and can't remember tomorrow because it hasn't happened yet. These circular answers tell her no truth. You may as well tell the child that the arrow of time points from the future to the past: You can remember what is going to happen yesterday because it really is going to happen but you can't remember tomorrow because it already has happened and is no longer real. Equally circular.

All of the basic laws of physics are time symmetric. There are esoteric particle events that are not, but these cannot possibly explain what is obvious to a three year old. Physicists have dredged the bottom of the barrel to find the reason for time asymmetry. They come up with the classical second law of thermodynamics: That entropy always increases. Prof. Feynman points out that entropy is also time symmetric because the laws on which it is based are time symmetric. To see entropy as an arrow of time we have to extend it to cosmological time scales. Is it credible that what is obvious to three year old is explained this way?

The asymmetry of time has a very stark and simple Quantum Mechanical explanation: The orthogonality of eigenstates. Then why do physicists dredge the bottom of the barrel of classical physics to explain time asymmetry? They are blinded by metaphysical prejudice. An example of this blinding is the Copenhagen Interpretation: It recognizes the reality of all eigenstates evolving deterministically until an observation is made. It even recognizes the reality of a photon encountering a beam splitter emerging in two orthogonal eigenstates. one is transmitted and the other is reflected. These paths may be meters apart and along which every atom of the apparatus potentially affects the calculation of amplitudes. These are summed for its eventual arrival at the interference screen at each point common to all paths. However, if you enter a state of awareness of one of these orthogonal eigenstates states and no other, only the state you are aware of is proclaimed to be real. It was selected as randomly chosen from the others by the

Uncertainty Principle to be the only one that is real. All the others, now not only are no longer needed, but urgently must be eradicated by the Copenhagen Vermin Squad. Thus we preserve the h=0,c=∞ world of one past one future, the hallmark of Newtonian Mechanics, the foundation stone of Physical Reality, homeland of the buggy whip.

When we talk of human consciousness, there is no possibility of erasing brain awareness because of the size and complexity of the system. If you spill a bottle of ink on the carpet, there is no hope of getting it all back into the bottle then forgetting it ever happened. However if awareness never gets more complicated than a particle in some state in some channel of the apparatus, and interference is destroyed by creating "which path" awareness, it is possible to restore interference by erasing the awareness. In addition to quantum eraser experiments described by Banning, a beautiful experiment illustrating this was done by S. P. Walborn, M. O. Terra Cunha, S. Pa´dua and C. H. Monken: *Double-slit quantum eraser* PHYSICAL REVIEW A, VOLUME 65, 033818 described in a web-page created as an assignment for PHY 566, taught by Prof. Luis Orozco at Stony Brook University in the fall semester of 2002.

There is no "Arrow of Space" in the sense that there is an "Arrow of time". If you are here, remember being here, you can go north, change your mind and come back here and find you are back where you remembered being. Or south, Or west, Or east, Or up, Or down, and come back here again. We are traveling through what we see as the physical world, streets, countries, planets, stars, galaxies. If we travel at the speed of light we are also traveling along time dimension of the outside world. But when we turn around and come back to the same place, we do not return to the past, but go further into the future. If a journey links A and B, the laws of physics describe the relationship between A an B and the points in between. They see no arrow. There a glaring asymmetry between A and B that is not accounted for by classical physics or the theory of relativity. Nor is it accounted for by the quantum mechanics if you superpose on it the claptrap of Copenhagen and von Neumann's Process I. It is obvious to our three year old: When she it at B, she can remember A; when she is at A, she cannot remember B. Everett stripped the malarkey off quantum math.

Suppose an evolving system is observed to be in an eigenstate Aa of the observing operator, at time T_0 and the quantum mechanics predicts that it will evolve into a superposition of eigenstates Ba, Bb at time T_1. At that time we observe it to be in Ba, say. Now the quantum mechanics predicts it will evolve at time T_2 into states Ca, Cb, Cc. . . If we experience

Ca and can remember Ba but we will not be able to remember its orthogonal state Bb, but can remember Aa. Is the converse true? If Aa remembers Ba it can also remember Bb but this would be a state of awareness that encompasses orthogonal eigenstates which is impossible. Thus Aa cannot remember the B states and Ba likewise cannot remember the C states. It is the structure of this hierarchy that imposes the "arrow of time" on the way experienced states are remembered. We cannot be in a state of awareness of more than one of states that are orthogonal eigenstates of the observation. The Copenhagen interpretation blinds us to this understanding because it collapses to non-reality any state we are not aware of. Now there is no quantum mechanical reason we can not remember the future so we have to look for a barrel to dredge to find one. This should give cause to a three year old to see Copenhagen as bunkum.

Thus memory of the future is debarred by orthogonality of eigenstates. The "arrow of time" is only a label. If all roads lead to the Big Bang, calling the "past" the road to the Big Bang and the "future" the road away it, only labels these directions. The real asymmetry is that the future cannot be remembered, the past can be. You can travel from the past to the future at the speed of light by sitting in a chair with your hands folded on your lap and your mouth full of teeth. The laws of physics tell us that the journey from the past to the future is the same as the journey from the future to the past. This is reflected in Newtonian mechanics as the algebraic sign of the time step in a numerical integration of the orbit of a comet. The same computer program can predict the future and postdict the past. Why waste time thinking about journeys from the Future to the Past when you cannot remember the Future when you get to the Past.

In principle we can remember the single chain of events that link us to the Big Bang, so it is meaningful to talk about **the** past because there is only one such chain, even if our present awareness is increasingly full of gaps in memory as we go back in time. The inference that there is therefore one path to *The Future* is pure metaphysics.

If we accept classical physics as Laplace did, there is only one path from the Big Bang to the end of time. We are merely observers of a passing parade of predetermined states of awareness and the idea of free will or being able to avoid calamity is a delusion. If misfortune befalls us, blame God.

The number of paths from any present moment into the future is uncountably infinite: It has the same infinitude as the irrational numbers which form a continuum even though each number like π is discrete.

These paths form a continuum in which each path is discrete and obedient to physical law.

Earlier, I described quantum eraser experiments in which an interferometer is constructed so as to determine which path the photon took. Doing this destroys interference. However an addition is made to the apparatus after the "which path" information is acquired. This addition irrevocably erases the "which path" information. Doing so restores interference. The quantum mechanics is time symmetric but awareness is not. How do we understand this without dredging the bottom of the classical thermodynamics barrel? Toss out the Copenhagen Interpretation for starters, and see that the present linear superposition of eigenstates of the causal system causes real states of consciousness of being aware of every eigenstate of the observed system. These do not bind in awareness. Each state of consciously experienced awareness is of the eigenstate that caused it, to the exclusion of all others which however remain real. This holds because of the orthogonality of these states. Running time in reverse causes the real (conscious) states of the linear superposition, all but one of which was collapsed by Copenhagen, to have been the cause of the state that was resolved onto the eigenvectors of the observation. This reversal is the quantum eraser that erases all memory of the future states. This restores to the quantum mechanics bidirectional causal linkage of classical mechanics without compromising the future-past asymmetry in awareness. Bidirectional causality does not imply the equivalence of cause and effect.

Those who respect Everett nevertheless may try to hang on to the physical world with a Many Worlds view of separate but non-interacting physical universes. Every quantum mechanical event spawns infinitely many of them. This was too much for John Archibald Wheeler to take on second thoughts after persuading his student Everett to put his ideas into a PhD thesis at Princeton. "It drags too much philosophical baggage into quantum mechanics" Wheeler said in effect. The problem isn't Everett. It is the "Physical World". It is necessary to stop believing it is real. It is not necessary to start believing consciousness is real because believing it is not is a real conscious experience. So there. The reality of consciousness does not play second fiddle to our beliefs. Infinitely many futures means infinitely many histories in consciousness not infinitely many physical universes. The validity of the conservation laws of physics and their symmetries along all possible paths in <u>consciousness</u> does not require belief in a physical world as reality. However, we will always be <u>aware</u> that the events of experience happen in some physical world in which conservation and symmetry laws obtain.

Lost in an infinite realm?

Creating the future in the image of the past is a metaphysical blunder making us seem lost on one trivial path in an infinite realm of future histories in consciousness.

This is delusion.

The truth is just the opposite: There are infinitely many futures that will remember you. Such memory may not include fame and glory, but in some way you will leave your footprints in the sands of time in all worlds in which such awareness is not in ruin. A hundred million years ago a dinosaur left her footprint in mud. That one small step for her was a giant leap for life on earth: Today, the number of worlds of consciousness that look at her footprint is infinite. In such worlds that cannot see it, her destiny is meaningless and may as well be in alternate physical universes.

Free will

There is no obvious and easy definition of free will. But there is an easy definition of what it is not. If one is prevented from making a decision for any reason be it coercion, domination, tyranny, carelessness, irresponsibility, parental authority, cautiousness, lack of sufficient information, one's will is not free. By default, we can define the basis of free will to be the ability to make a decision. Doing so is no guarantee that ones decision will be wise. It may be stupid or the arrogant decision of a tyrant. One can define war as the attempt to impose ones will on others by force and their willingness to use force to prevent it.

What should be understood as a landmark event of history emerged by the 20th century. The ability to move decision making outside the human brain. Machines that did this were called automatic. The first concept of a purely decision making machine was due to the mathematician Turing, but it was only a conceptual machine to place mathematical reasoning outside the human brain where it could be seen. The first practical machine of pure decision making reasoning was invented by von Neumann. As the mathematical guru of the quantum mechanics, he was placed in charge of the mathematics lab at Los Alamos during the development of nuclear bombs in WWII. The lab was equipped with IBM punch card programmable calculators. These machines were not capable of making decisions. A program was a deck of cards, one card per instruction. After running a program deck of cards through the card reader, a physicist would look at the printed results and decide what program to run next. If the needed program did not exist, von Neumann would see to it that it was written, punched and tested. He understood both the physics and the math.

Too late for WWII, but thinking how the punch card calculator could be improved, von Neumann invented a machine in which all needed programs would be stored in a single electronic memory. The machine is given the capability of being programmed to make a decision as to which program to run next based on the results of the earlier calculations. What is remarkable is that to have such enormous capability such a machine need have only one simple decision making hardware instruction. This could be done with a branch instruction that would go to memory location A for its next program if a testable condition was true or default to program B otherwise. The testable condition could be as simple as the arithmetic register overflow. It may seem that such limited capability would give the machine very limited decision making capability: However, it was von Neumann's genius to realize that any humanly understandable decision to choose between two alternatives could be programmed to make the adder overflow in one case and not overflow in the other. The branches can branch and they can branch and so on, making complex choices among many possible alternatives. All modern electronic computers are von Neumann machines. Not many decades after this invention, a programmed computer defeated the world champion at chess. Present day computers have many conditional branch hardware instructions that test diverse criteria, but this plurality does not increase their power, it only makes them more convenient to program.

It may seem that a von Neumann machine has no free will because it will always compute exactly the same results given the same input data. In fact, if it didn't do this it would be broken. What this reasoning fails to take into account is that the machines that defeated Kasparov did not have advance knowledge of his moves. Otherwise the machine could have looked up a humanly predetermined response in a table of all possible moves. Kasparov moves came into the machine from the outside world of which it had no prior knowledge. His moves may never before have occurred in any game of chess. Given any move he made, the machine was free to decide what response was best. The art of programming such a machine is that the programmer with no knowledge of the moves the machines opponent will make must give the machine a power to select the best alternative the programmer can think of, which may not be the best possible. This particular decisions made in the game with Kasparov, not foreseen by the programmer, may be extremely complex, involving the execution of conditional branch instructions testing different criteria millions of times for every move Kasparov made. We cannot claim to know that the human brain behaves any differently in the exercise of free will. So as not to appear predictable, a von Neumann machine can be programmed to behave randomly when it is faced with a decision between two equal alternatives. It can count the number of Geiger counter pulses in a second and do one thing if it is even and the other if it is odd. The machine is going outside itself, unless the radioactive sample is defined to be "inside the machine". In either case, randomizing the machines response does not necessarily give it greater power unless such behavior confuses the opponent.

A wise man whose will is free always makes the same decision given the same facts. (Only wise women have greater freedom!) The proof is seen by looking at the converse: The wise man's decision is countermanded by that of a dominant fool, the wise man's will is not free. There in no inconsistency between free will and reality being deterministic. An isolated process may be deterministic and still have freedom of will in the way it deals with the outside world when the isolation is interrupted by coupling with the outside world without coercion. But what if the outside world is itself an isolated system. This infinite regression was discussed by Everett who concluded that it makes sense to talk about the state vector of the entire universe. We have to ask: What is the ultimate boundary outside of which we cannot go? I will define three such boundaries called god2, god1 and god0.

- god2 is the physical universe of Classical Physics. There are no decisions in this world. Only one path from the Big Bang to the end of time is real. Kasparov and those rooting for Deep Blue are fooling themselves that anyone but god2 is playing the game, and he has a one track mind that considers no alternative.
- god1 is the captain of the Copenhagen team. The game is the Copenhagen Defense. But, not even god1 is playing the game. The toss of his dice determine every real event – Kasparov, Deep Blue and god1 himself are powerless pawns in a stochastic gambit sacrificing causality only to have it reappear as a superluminal apparition annihilating metaphysically undesirable realms of inconvenient reality. Or something.
- God0 A infinite continuum of deterministic paths in consciousness from the Big Bang to the end of time. Where sufficiently intelligent beings are found they will be in states of consciously experienced awareness of their interaction in spatial and temporal

proximity that follows a single path back to the Big Bang. They will have no experienced awareness of the infinitude of paths that lead to their future. But they may have accurate models of the future that can be used to make decisions as to which paths to avoid and choose where this is possible. This continuum is not inside space time. It is eternal, in the sense of being outside time. The reality of consciousness is deterministic but the journeys in awareness one can make through it are not coerced, even by God.

Suppose you are debating whether to take two courses of action A or B. There are pros and cons for each and you are teetering on the brink of indecision. Instead of waiting for quantum events in your brain to select a course, you turn on your beam splitter and let a photon make the choice by being transmitted in which case its A, or reflected in which case its B. Shades of Schrödinger's cat. You press the button and now you know, its B. Is this free will? The answer is yes: You and the photon are the same conscious self. You had no free will in deciding if the choice should have been C or D. There will be another real path in which A was chosen and that conscious self will be you in consciousness, but A and B are not in the same state of awareness. Each will follow different paths in awareness which were not chosen under coercion. The photon didn't make you do it. You and the photon are one. Both choices provide a happy home in reality for defending the respective pros and down playing the respective cons. In the course of time you may regret your choice when you find that powers beyond your control did not share your optimism. Your will is free because you made a decision you were not forced to make. Only the impossible is prohibited. Blaming the photon is regressing to classical metaphysics: a cop out. If you made an unfortunate choice, you will not leave it to chance next time. Everett effectively tells us to take Yogi Berra's advice: "If you come to a fork in the road, take it!" If the laws of physics chart two roads A and B, you will take both as real journeys in consciousness, but neither will be aware of the other. You freely took A because you were not coerced. You freely took B because you were not coerced. Each will lead to its own destiny in awareness. If your choice B becomes regrettable, you can choose a path that blocks it before you get to the next fork.

Saying that there is nothing we can do about misfortune is not true if we can identify causal events that lead to, or prevent, misfortune. If you know you family may be killed by a tornado it is your fault if they are, if you built a swimming pool instead of a storm cellar. It is your fault if they are killed by an earthquake, if you built a tennis court instead of reinforcing the cripple walls of your house. This is not a lecture on obviousity, but a statement of what may not be obvious: The above would not be true if classical physics is correct or if the Copenhagen Interpretation is correct. Everett's Relative States interpretation seen in the light of what has been said above about consciousness and awareness tells us that if we make unfortunate choices we will consciously experience misfortune. We have freedom to make choices in awareness but not consciousness. This accords with common sense and intuition which only lead us astray when we interpret them in the light of metaphysical delusion. Belief in freedom of will, embodied as it is in law, would be baseless if the Copenhagen Interpretation, or classical physics, were correct.

Religious cop-out in Christianity is to attribute avoidable misfortune to God. "God took my child when the tornado killed her". Islamic cop-out under the same circumstances is "If it is written that we should die, overtaking on an outside curve cannot change what the moving finger will write". Yes, it is written that the evil of irresponsible stupidity is rewarded by avoidable tragedy.

There is an ancient argument that God cannot be both omniscient and omnipotent: If God knows what is going to happen, God is powerless to prevent it. If God decides to prevent it, God could not have known that it was going to happen. There is a fly in the ointment of this argument. The fly is you. You are the freedom of will of God. This is not good news for those who freely choose the wide path to destruction and not the narrow path to life. Of course, you could sincerely appeal to God's omniscience for guidance, but that won't work if you do not believe that God is listening, or sympathetic but incapable of responding.

Summary

Physical Objects

These are the foundation stones of Physical Reality. You are a physical object, so is your car, reading glasses, and toothbrush. So are the pebbles in the brook. They occupy a definite region of space and have a definite state of motion at all times. They are made out of atoms and molecules. Atoms are made out of elementary particles: electrons, photons, quarks, and gluons.

We can make two inferences:
1. Elementary particles are physical objects too, which explains why everything made of them are also physical objects.
2. Physical objects are mathematical approximations to configurations of elementary particles; reality created in the image of these approximations is illusion.

The first step in understanding physical objects is to look at space itself. The quantum mathematics describes electrons and quarks as having a definite intrinsic angular momentum which has only two "spin" states. They are described by 2x2 matrices, but three of them are needed. Feynman shows us how to derive these matrices from nothing + the quantum principle of superposition. The direction of stars in the sky have only two states given by two angles astronomers call α, δ when based on the equatorial reference plane plane. A surveyors theodolite uses different angles az,el based on the horizon plane, but it can also look at stars with only these two angles. These two systems are related by linear algebraic equations: exactly the three given in Appendix E, represented by a 3x3 matrix, which represents three dimensions, not two. The six vectors of this matrix all obey Pythagoras' theorem of the sum of the squares of the elements equaling one so each vector only has two degrees of freedom. What about time? Minkowski shows us that Pythagoras' theorem $s^2 = x^2+y^2+z^2-t^2$ where x,y,z is space and t is time describes the special theory of relativity, the birthplace of quantum mechanics. Dirac and Einstein show us that this means electrons must have spin. So the circle closes and it is all quantum mechanical.

We now ask if an electron can be seen as a physical object. If its state of motion is known its position in space is is unknowable, or it fills all of space, however you want to see it. If we know its position we cannot know where it will be seen next. If you have some electrons, you can't tell one from another. If you see a photon from a star at a definite point on your retina, its wave-front just before it entered your eye covered thousands of square light years of space. If you see two photons from two close stars, you cannot tell which photon came from which star. An energetic photon can morph into an electron and positron. These games are played in quantum mechanical space. They do not look like the foundation stones of Physical Objects.

It is fair to ask if atoms can be seen as physical objects. Under certain conditions yes. If you accelerate an electron to 10,000ev, it behaves very much like a charged bullet. At 50,000ev relativistic effects set in and undermine the interpretation that it as intrinsic attributes of immutable physical objects. The truth is that in that energy range we can see, with considerable accuracy, the electron as a physical object. A proton even more so because its rest mass is 2000 times greater. But it is only an approximation.

Suppose a slow electron and proton get married and make a hydrogen atom. Now the electron can take on definite states of angular momentum. But its position is completely indeterminate in angle, but because it is not everywhere we can put the atom in a small box. Not too small or the electron will have a finite Born probability to be outside the box. Now two hydrogen atoms can get married and make a hydrogen H_2 molecule. A step in the direction of becoming a physical object. Not really: When is wobbles or spins it emits and absorbs light following quantum mechanical rules. Now H_2 and O get married and make a water molecule H_2O. Chemists have known for centuries how to make molecules out of atoms well before the days of quantum mechanics. Lots of H_2O frozen will make a snow crystal. Finally a physical object! No, not really; we can only understand the van der Waals forces that hold the crystal together using quantum mechanics. Lets make people instead. We know they are physical objects, don't we?. Get some H, C, O and N atoms and make nucleotide bases. Make some deoxyribose sugar molecules. They will all fit in boxes. Now make DNA molecules the way Crick and Watson did using tin models which we know are physical objects, aren't they?. The tin models they used could be seen as physical object, but had to be made with bonding angles computed with quantum math. Well then, lets make a car engine; we know that is a physical object, of course. Mechanics can do this without taking a single quantum mechanics course. Start with nuts and bolts. A screw thread is a helical ridge and a nut a helical groove that intertwine. No quantum mechanics, just combinatorial geometry: Just don't mix national fine thread pitches with national coarse or metric. They are mathematically exact objects permanent in time always occupying a definite region of three dimensional space. Finally we are in Physical Reality, the homeland of physical objects.

We now know how to set the bounds of the physical universe:
- Set Planck's constant h = 0;
- Set the speed of light c = ∞;
- The finishing touch. Set the curvature of the earth = 0. There!

Einstein said that it is a miracle that anything is comprehensible: Nesting within each model there is another that is simpler. The simplest is comprehensible to animals and even insects. They live in the physical world, homeland of physical objects. There is a price they have to pay to be denizens of this simple world: They will never understand black holes and Mandel interferometers. That may be a blessing.

It was natural for the founding fathers of the quantum mechanics to believe that its mathematics described the physical world which one could do with the addition of metaphysical artifice that Einstein rejected. He changed the quantum math in a way that removed the metaphysics but retreated into a theory that made predictions shown, only after his time, to be wrong. To this day, most physicists have a death grip on physical reality, refusing to believe that there is no such place, refusing to believe that there are no such things as physical objects, ever blind to the reality of consciousness that has been the only contact they have ever had with reality every waking second of their lives. Why cannot they accept consciousness as reality? They cannot accept consciousness because it is not a physical object: It does not exist. Whatever is real <u>must exist</u> they believe. It took life on earth 4,000,000,000 years to reach the point of understanding that this belief is delusion. Something does exist: Mathematical truth, but you can not make it out of physical objects. Sorry, Bertrand.

To animals and some humans the Physical World is God. Newton was not one of them: He believed that God created the Physical World, imbuing it with certain laws and he gave us those laws. In the twentieth century, we learned that these laws are only approximations to a deeper truth.

We are forced to accept the reality of consciousness as a miracle: We can never know why it is real, because we only experience it, but never can observe it: It does not exist. It is not reasonable to doubt the experience of consciousness of any being or thing. However, one may doubt <u>awareness</u> of consciousness in any brain but those of chimpanzees, dolphins, whales and homo sapiens, and chimpanzees may be a stretch. There is a simple reason that human awareness is never manifest except in consciousness: It is not because consciousness is a spin-off of something else, imagined to be the "real" reality. It is because there is no other reality. We do not understand now, but may in the future be able to understand:

- How objectively describable neural firing patterns correlate with subjective consciousness;
- The grand scheme of awareness of the human brain and hence the design of much simpler brains such as that of the honey bee;
- The role of the subjective content of consciousness in evolution.

With each fragmentary advance after each advance we make in such understanding we will reward ourselves with Nobel prize after Nobel prize. What we will be understanding is not the spin-off of "accidental" collocations of atoms. In von Neumann's Process II, there are no accidents. We will be understanding God's design of the temple of life, which includes the role played by subjective conscious states which are miraculous and unobservable, or if it pleases you to make the distinction, by evolution. May I suggest that the distinction between these views be laid to rest beside the grave of the Physical Universe and the urn of ashes of the debate: How many angels can dance on the point of a needle?

The Physical World and the quantum Mechanics

We learn at an early age that the world outside our heads is much bigger that the one inside. We are fallible. The external world is not. We cannot remember where we left our keys. The outside world never forgets. It is the rock of ages, immutable in space and time, immutable in mass/energy proclaiming its ponderable solidarity.

Until 1927.

1. When a photon comes to a beam splitter we are forced to compute its amplitude to take **both** paths to explain interference.

2. If we look in both paths, the photon is seen in one but not the other.

But it took both paths through the beam splitter. What happened?

Still believing in physical reality in 1927, in 2. above we must believe that the path observed can be the only one that can be allowed to be real. Others, real to that point to account for interference, now must be destroyed instantly everywhere. No, we say to ourselves, we cannot allow the physical universe to die. If what Alice saw was caused by the photon passing the beam splitter, superluminal causality has to make Bob see the same thing. Darn it, even light itself is not fast enough to do the job. But it <u>has</u> to be done or the Physical World will go to hell in a hand-basket. Wait a minute! Who saw it first, Alice or Bob? Einstein says some observers in motion with respect to Alice and Bob thinks she did, other observers think Bob did. Who is causing what? OK then, lets give up causality too.

Either these statements are hogwash or the "Physical World" is real.

If we no longer have the ultimate reality of the physical universe to believe in, what else is there? It is very tempting to say, OK the physical world is an approximation, but reality is very much like it. OK, electrons are not little hard balls of charged matter, always somewhere and doing something, but they are very much like that. When reality is understood to be consciousness, there is absolutely no resemblance between reality and the physical world. One cannot imagine two things more starkly different. But, one says, I can go back to the same place and find the same thing is still there, which means the old physical reality is still there. One can go <u>always</u> back to the same place in established coordinates. Is that the same place in physical reality? Go back to longitude 0° 00' 00.000": a white stripe on the floor of the meridian circle telescope at Greenwich, and check the longitude with your GPS. Now go to the same place in GPC coordinates. You will not have the conscious experience of being in a meridian circle building with a white stripe in the floor. It is not the same place. Sorry, that the best astronomers can do.

I grew up on a gold mine where a forest of black 100 foot high headgears and 200 foot high yellow tailing dumps extended along the Rand from horizon to horizon. I look at the same place on Google Earth and it is all gone. I can find where my house once stood near the intersection of a derelict remnants of a weed covered street and a similar street going skew to Joan Brinkley's house. If I were blindfolded and put down there and asked to say where I was, not having seen Google Earth, not only could I not say where I was, I could not even tell you what continent I was on. Conscious states are correlated and what they have in common is what we can represent with $h=0$, $c=\infty$ approximation:

The reality of consciousness explains the paradoxes of relativity and the quantum mechanics. We do not have to try believing in anything. Consciousness is real no matter what we believe. All we have to do is stop believing in illusion. Our built in model of physical reality allows us to make quick and sensible decisions, like where we left our car keys or whether it is safe to stop before the light turns red. However, it is an intellectual blunder to believe that, basically, we live in the physical world, but sometimes there are quantum and relativistic effects here and there. As best we can understand reality it is <u>all</u> quantum, it is <u>all</u> relativistic. To believe otherwise is akin to believing that, basically, the earth is flat everywhere except sometimes there is a little curvature here and there put there to assuage eggheads.

In a lecture by Prof. Susskind in his Stanford series *The Theoretical Minimum*, he describes the state of a photon interacting with an atom as possibly being in a linear superposition of existing: $|1\rangle$ and not existing: $|0\rangle$. We could write this as photon state as:

$$\alpha|1\rangle + \beta|0\rangle, \text{ where } \alpha^2 + \beta^2 = 1.$$

A Physical World model cannot represent this state because simultaneously existing and not existing is impossible. This is a much starker inconsistency than a photon simultaneously existing in two contradictory states in an interferometer, like being transmitted by a beam splitter while at the same time the same photon is reflected. In Everett's formulation of the quantum mechanics we cannot encompass existing and not existing in a single state of awareness, again for reason of consistency, but we can imagine each as being equally real. If it is that simple, why was Everett's formulation not universally adopted in 1957? This rhetorical question is answered thus: By avoiding use of the word *reality*, he chose not to cross the frontier between science and philosophy, leaving the world blind for sixty years to the truth he discovered. Our blindness is our fault, not his. He did his job in giving us the quantum theory of awareness. The lesson for us is that we cannot turn a blind eye to reality: It is consciousness.

Von Neumann did not turn a blind eye to consciousness, nor did Everett in showing consistency between his formulation and the doctrine of psycho-physical parallelism. Everett showed von Neumann's Process I to be metaphysics. The fatal lozenge that Everett swallowed was von Neumann's proclamation that consciousness is in the Physical World, which makes inconsistency in consciousness impossible. The antidote to von Neumann's poison is the realization of the converse: The Physical World is in consciousness. The Physical World exists as a now debunked mathematical theory highly accurate in many domains of experience. Reality created in its image is delusion. Consciousness does not exist, so it may take us on journeys through real experiences of inconsistent states of awareness which do not bind in a single state of awareness. We need not turn a blind eye to reality. Open our eyes to see it, our ears to hear it, our nose to smell it, our mouth to taste it, reach out to feel it. Consciousness is not a theory: It cannot be because it does not exist. Consciousness is reality with a structure represented by a highly geometric theory: In its simplest form we call it Hilbert space. We can never understand why nothingness is the real substance of consciousness. Following the twentieth century we can understand how

God created the physical world out of nothingness. The quantum mechanics is the Zen logic of creating something out of nothing. If absolutely nothing exits, nor ever has nor ever will be, you will have the conscious experience of seeing pebbles in the brook, which you can see without consciousness having to exist.

Feynman diagrams

The quantum mechanics grew out of attempts to understand the absorption and emission of photons by hydrogen atoms when its electron changed its energy states. In same time frame, experiments with cosmic rays, radioactive decay and collisions of artificially accelerated particles seen in cloud chambers which made particle tracks visible, showed particles morphing into one another. These studies led in the late 20th century to the understanding that all particles are related to two families each of six particles, leptons and quarks which exist in two mirror image factions called antiparticles. Forces between them are mitigated by a family of bosons. These so called particles are classified by group theoretic structures. It is not completely understood why they are what they are. It would not have been surprising if the simple quantum mechanics of 1926 and 1927 was only the beginning of a much deeper and more complex theory. That was not the road history took. Instead it turned out that the theory of Heisenberg, Schrödinger and Dirac was the basic story of the way everything works that was not undermined by later advances in quantum electrodynamics and chromodynamics.

Feynman diagrams look like particle collision event pictures, are schemes used describe the integrals needed to calculate the quantum mechanical amplitude for a particle interaction event to occur. Lines represent propagating fermions and wiggly lines represent propagating bosons. Up page depicts time and across the page space. Vertexes represent interactions. Particles going backward in time are antiparticles. A collision event mapped in a cloud, bubble or spark chamber or equivalent device of which now there are many, may require more than one Feynman diagram, even many, to represent each way the event can occur. The amplitudes for all must be calculated and added to get the amplitude for the event to occur. Each Feynman diagram depicts not one but a continuous infinitude of ways that the event can happen which must be summed over by the methods of the integral calculus. They can be thought of as visual aids to remind the physicist of which integrals need to be evaluated to represent the amplitude of an observed event.

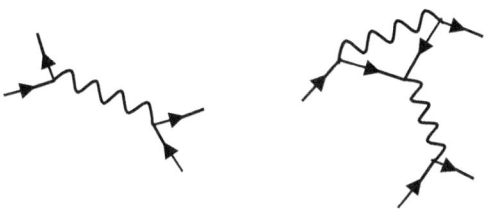

To do the necessary calculations to represent the interference fringes seen in an interferometer, or the image formation in an astronomical telescope, one does not need Feynman diagrams. Nevertheless, the necessary integrals must be evaluated for every way a photon, from say a star, can reach each point in the image plane including all possible paths a photon can take through the interferometer or telescope objective lens.

Each diagram can be thought of a something happening in the "physical world" but it takes the superposition of infinitely many of these physical world models to represent the real world of quantum mechanics.

If you accept the premise of this book, there is no such place as the "physical world" in reality. But it does exist mathematically as a simplifying approximation and can make very accurate predictions, like eclipses of the sun and moon. Feynman diagrams can make even more accurate representations of quantum reality:

EPR and Everett

Einstein, Podolsky and Rosen were on the right path when they rejected the notion that Alice's activities do not causally alter what Bob sees, however remote he may be. They argued that the correlation between what Alice and Bob saw was predetermined. They were exactly right, but along that path they fell into a pothole of physical reality and tried to get out by arguing that the quantum mechanics was an incomplete theory by patching onto it hidden parameter properties of particles. It took 50 years to prove this hypothesis wrong. There is a much simpler explanation: Every time a photon is emitted (supposing their experiment is done with photon polarization), <u>exactly the same thing happens</u>. If, with respect to their parallel polarizers, the photon is in a superposition of two orthogonal eigenstates, the conscious experience of being in states of awareness caused by each eigenstate will be real, but will not bind in awareness whether or not enough time has elapsed for Alice and Bob to compare notes. If the experiment is repeated over and over <u>exactly the same thing happens every time.</u>   If and when they do compare notes, they will always agree because they whatever they compare will be caused by the same eigenstate of the observation.   It is baseless to worry that Alice being aware that the photon was transmitted by the beam splitter will run afoul of Bob's awareness that it was reflected: These states of awareness ate caused by orthogonal eigenstates of the beam splitter which do not bind conscious experience of both.   Nor do we have to worry that there are millions of conscious souls to be divided into two herds, the reflected and the transmitted. There is only one: You.

We can accept Everett's formulation and hang onto the physical world by believing the photon is both transmitted and reflected, as the possibility of interference requires, two physical universes now are real. In one the photon will be reflected which both Alice and Bob will be aware of whether or not they subsequently send notes to one another confirming their agreement. Likewise there will be another universe in which Alice and Bob are replicated where they agree that the photon was transmitted.   This view gets rid of randomness and superluminal causality.   Getting rid of the physical world and replacing it with consciousness gets rid of multiple universes.

If it is that simple, why did the quantum pioneers not see what it was: They were no dummies.   It is necessary to distinguish between consciousness and awareness. One binds the other, one is scientifically observable, one is not.   The distinction cannot be made unless the Samkhya Paradox is resolved which requires accepting the universality of conscious self-hood. It was not the inability to comprehend simple truth that stymied them. It was their inability to give up a metaphysical belief: That they were not voyagers in consciousness, but travelers among physical objects in an ultimately real physical world.   I doubt that it crossed their mind to question this. Ten years from the time I write, the Copenhagen Interpretation will be a century old.   It remains the predominant interpretation of the quantum mechanics.   Einstein and Schrödinger detested it. Einstein: "I am convinced that He (God) does not play dice." Schrödinger, writing about the probability interpretation of quantum mechanics, said: "I don't like it, and I'm sorry I ever had anything to do with it." He died a few years after Everett's thesis which otherwise may have given him cause to withdraw his statement.

The frying-pan

Why did so many brilliant people sit sizzling in the frying-pan of Copenhagen for so long? No person, not even Einstein, will jump out of it unless they can see where they are going to land. Einstein did jump after preparing, with Podolsky and Rosen, a spot to jump to. After Bell and Aspect, it was to no avail and went up in smoke after Einstein's time.

After Everett, one could accept multiple realities only by accepting multiple physical universes bursting out of each quantum event. There is a simpler way: dump the physical universe. Berkeley did and he didn't need all the clues we have been given.

Richard Feynman's words, *But how can it be like that?*, laid to rest by Everett, need to be exhumed with the understanding that the word **it** in bold type refers not to the quantum mechanics but to consciousness.

We can imagine Sperry and Gazzaniga's experiments extended into the white matter axons in each hemisphere of the brain to elucidate in more and more detail how the brain works until it reaches the cortical neurons. Finally we have a complete understanding of brain function. Imagine that physicists at the same time finally have a theory of everything – a theory allowing them to compute the 16 constants that now have to be observed – a theory that unites quantum mechanics and general relativity. Imagine we can understand the neuron-chemistry and physics and relate that to the theory of elementary particles. Will we finally explain consciousness? If the premise of this epistle is correct, we will have a complete model of awareness but will still have no clue why anything is real. We will stare through consciousness, blind to its reality, looking for reality beyond it created in the image of our theories about it. There is a place where you can find incontrovertible proof that consciousness is real, and you do not have to do anything or go anywhere. That place is the fly in Schrödinger's ointment. That place is you. That thou art. *Tat tvam asi.*

You can avoid unfortunate consciousness awareness but you cannot escape from consciousness. There is Absolutely Nothing to escape from. Everything science has taught us is the structure of consciousness. We need to learn the lesson of history and not try to create reality in the image of our theories. This is not new advice: Remember this?: "Create no graven image of anything on heaven or earth." Note the earth part. All theories are images. We can take the advice of not creating reality in their image.

The chess game of psycho-physical parallelism leaves only two kings on the board and the stalemate is the boundary. The mating move: Remove the King of the Physical World from the board. The rules of chess have been too kind to him.

Niels Bohr

He stated that the job of science is not to say what "is", but, in effect, how things work. Except for a semantic collision between "existence" (*is* being a conjugation of *to exist*) and "reality" nothing in this epistle disagrees with Bohr's view. However it is scientifically important to say what "is not". This must be done, not to contradict formal philosophy, but to contradict the naive philosophy of scientists. Everett said this: *However, when a theory is highly successful and becomes firmly established, the model tends to become identified with "reality" itself, and the model nature of the theory becomes obscured. The rise of classical physics offers an excellent example of this process. The constructs of classical physics are just as much fictions of our own minds as those of any other theory; we simply have a great deal more confidence in them.* (see p134 of the DeWitt/Graham book: *The Many-Worlds Interpretation of Quantum Mechanics*): This is the clearest statement I have seen in print that illustrates how gut level metaphysical belief, seen as reality, may become blessed with the holy water of science and, in the case Everett refereed to, by a theory known to be wrong.

Everett extended this view to cover metaphysical add-ons to the quantum mechanics, particularly to the Copenhagen imposition on reality

- that it is inherently unpredictable
- observation collapses wave-functions

Both are needed to protect preconceived metaphysical views of reality.

To succeed, science does not need to accept consciousness as reality. The task of science is not to explain reality but to model awareness. The merit of accepting consciousness as reality is that it leaves no room for metaphysical "realities" to get their foot in the door. There is no scientific proof of the reality of consciousness. Consciousness is not metaphysical because there is, nevertheless, an experiment that proves it is real: You. You can do good science as a solipsist believing only your consciousness is real and the rest of us are clever imitations of you: Turing Imitation Game automatons bereft of consciousness. If confronted by the paradox of the Samkhya philosophers, just shrug your shoulders, and look the other way.

It and you

Even now in the 21st century, if you read books on physics or listen to any of the many online lectures, you will see the extent to which the physical world model entraps people's minds. If one is talking about a particle like a photon or electron, the observer is seen as objective bystander observing the particle as and external physical object. We talk about its momentum, its position as pointer readings in an imaginary physical world model. If we make definite observations we say: "Ah hah. Tried to hide from me behind your veil of uncertainty, didn't you? Well I gotcha now! All those other places you fooled me you could be? Well you'll never get back to them: Eh Hey! They're goners! Collapsed!"

This metaphysical entrenchment turns a blind eye to the lessons of he quantum mechanics. If we are interacting with a particle, there is only one wave-function that describes both it and describes you. You and the particle are one. What you are doing is an operation described by an Hermitian operator. It does not tell you, an imagined impartial observer, what state the particle is in, seen as physical object external to you. As a result of the interaction, it tells you the state in which <u>you</u> are left, as a eigenstate of <u>your</u> observation operator or possibly in a linear superposition of them. Even when your mechanism of awareness and the particle are not interacting, you and it can be described by a single product state wave function of the particle on one hand and your mechanism of awareness on the other. If the latter is evolving in time, you can drop the particle, and metaphysically interpret it as a separate physical object. If you had the omniscience of God, you only need one wave-function to model your evolution and anything else: The wave function of the entire universe. You would need the omnipotence of God to do the calculations, but could make life a lot easier for yourself by leaving stuff off, like α Centauri and the Eiffel tower, if you aren't going to Paris any time soon.

In Lecture 8 of the ***Theoretical Minimum*** online series of Prof. Leonard Susskind, in answer to a question I could not hear, he said: "All I know that is true is this." pointing at the math equations he had written on the board. "There is a lot of c**p out there, a whole mountain of it. What I am trying to do is to cut through the c**p." I could not agree more except in this book I have to be more polite, because Australians may be reading it, and agree instead that there is a huge crock of malarkey out there. In the same lecture he spoke of distant Alice on α Centauri; her state collapsing instantly following a Bobs observation in Palo Alto. In many places in his series he emphasizes the random nature of the quantum mechanics. A lot of ~~c**p~~, sorry, malarkey was put in the crock by Prof. Susskind. He seems uncomfortable talking about reality: If one can accept consciousness as reality, one does not have to talk about reality in science. The reality of consciousness is a very different place than the pseudo-reality of the physical world. The infinitude of histories in consciousness that has been going on since the Big Bang and is infinitely vaster than the single history of the physical world of Laplace or the single random rocky Copenhagen road. When we look to the past we see the single physical universe of our awareness as an infinitesimal place of all that is real since the Big Bang. The view to the future is completely the opposite: the physical universe of our awareness is the gateway to infinitely many worlds of destiny. In some way, each will remember us.

The three most important things to know about Prof. Susskind in order of importance are: He

1. is a configuration of leptons, quarks and bosons;
2. experiences conscious;
3. understands the math of physics.

The three items above are scientifically demonstrable. The most difficult is item 2 which requires you and Prof. Susskind to see a good neurosurgeon to have your visual cortex exchanged with his. If you come out of recovery and pass every scientific test of visual awareness but are completely blind you will know Prof. Susskind's brain does not experience consciousness. The result of this experiment is such a foregone conclusion that it need not be done: Just a well, because it is expensive and not covered by your medical insurance plan. Whatever you and Prof. Susskind decide to do, whomever comes out of it being called Prof. Susskind, could cut a lot more c**p out of physics by using Occam's razor to excise the physical word from quantum mechanics and dumping it in the malarkey crock: Doing so would drag in its appendages: random causality, collapsing wave functions, superluminal causality; in-deterministic observations without harming the non-commutativity of complementary operators.

Historical note: Malarkey crocks are real. At Lick Observatory all books donated to the library had to be kept because it was a state institution supported by tax dollars. Books advancing science were put on the shelves. There was an abundance of books written by scientific crackpots. In the library there was a large and nearly full locker of these books, called the "Crank Case". I am going to donate this book to the Lick library. Whenever Prof. Susskind visits Lick, he will check the Crank Case to make sure my book is there. My philosophy is Wabi-Sabi. (Also my dog's name): The perfect Malarkey crock must include a little imperfection.

Philosophy of life

It may seem that Everett's infinitely many paths in awareness trivializes the one you are on. One may be sobered by the consideration that in consciousness it does not matter what happens, it is going to happen to you. *Tat tvam asi.* In this light, the selection of one's path in awareness becomes all the more important than that afforded by the belief that the path one is on is the only reality and whatever goes wrong can be blamed on the toss of God's dice in the gaming saloons of Copenhagen, if not blamed on the determinism of Laplace. The focus of all religion is on path selection whether it be non-theistic like the teachings of Buddha or theistic like the teachings of Jesus.

It is not a return to the worship of The Physical World as God to realize that the prediction by astronomers of the coming eclipses of the sun and moon at split second times given by an atomic clock are true of all possible quantum mechanical worlds. We should by now have learned the lesson that the physical world is a very bad way of trying to understand quantum mechanics, but it is an excellent way of understanding that there will always be a Statue of Liberty, an Eiffel tower, and of course, there will always be an England, especially after Brexit. At the other extreme, the infinitude of paths taken by the quarks and gluons in every nucleus is of no importance to us because we cannot distinguish them. Nor are the times of decay of radioactive nuclei: 5000 atoms of potassium decay every second in your body and not knowing which ones did a second ago does not matter. Hurricane and tornado seasons will happen as usual: The only important thing we need is adequate warning by the weather bureau. We do not need a list of the infinitely many places tornadoes will in reality strike: we will not be aware of being in almost all of them. But you will consciously experience all of them, in one domain of awareness or another. If a tornado runs into you, do not complain: "The Uncertainty Principle didn't allow me to know that it was going to happen". Throw the Uncertainty Principle in the malarkey crock and listen to Everett: If a tornado is happening to you, it <u>was absolutely certain that it had to happen</u>. You have to prepare for <u>every</u> tornado that can possibly be real and there are infinitely many of them. And <u>you will</u> experience <u>all</u> of them in consciousness even though you will rarely experience one in awareness. That is the only way you can be responsible to yourself and your family. If absolute safety is not economically feasible, you have to be prepared to die in some destinies that you may live in others. Or move out of tornado country. But do not move into earthquake country, hurricane country, flood country, lightening country, volcano country, tsunami country, endemic disease country, or any country without country music.

Unless we worship The Physical World as an idol, there is very good reason to trust intuition and common sense particularly as they relate to free will: They were forged in experience of millennia, as my dad even believed this about the Ten Commandments which he obeyed while refusing to believe them to be of divine origin.

What happened to Schrödinger's cat

The paradox of the quantum mechanics was illustrated by Schrödinger's cat in a box being in a linear superposition of being alive and dead. Until you open the box. If a radioactive atom decays within some time period, a diabolic mechanism will kill the cat. Otherwise the cat lives.

The source of this paradox is the inability of most people to see the cat and the atoms it is made of as anything other than physical objects in ordinary space. However, as we shall see below, we cannot view the radioactive atom that way. We must look at it quantum mechanically. Schrödinger's clever parable forced us to look at he cat quantum mechanically too. Most people are not willing or able to do so.

An example of $h=0; c=\infty$ thinking is a radioactive atom mind trap, discussed at length online: You have a iodine atom, I_{131}, decays to xenon, Xe_{131}, with a half life of 8 days. Every time you look at I_{131} it still has a half life of 8 days. So like the watched kettle that never boils, watching it indefinitely prolongs its life. If you believe this, you are trapped in the Physical World, homeland of the Copenhagen Interpretation. Instead of dredging metaphysical hokum out of gut level intuition, we need to look at what quantum math is trying to teach us:

The state vector of this atom is $\alpha.|I_{131}\rangle + \beta.|Xe_{131}\rangle$, a linear superposition of two eigenvectors. $|I_{131}\rangle$ is an eigenvector Hilbert Space pointing at the **reality** of the atom being I_{131}. It is not a vector in space or time. The same is true that $|Xe_{131}\rangle$ points at the reality of the atom being xenon. However, as soon as I say *reality*, Dr. Johnson will walk up and say, "I know <u>exactly</u> what you mean by *reality*. It is the physical rock I kicked, and that is where they found your real physical world atom I_{131}. I say "No! No! Dr. Johnson, the physical world is not real, only consciousness is". John von Neumann hears this and says: "Johnson is right, Robert. But you are right too about consciousness. You see, consciousness is **in** the physical world".

Instead, if I use the word *consciousness* instead of *reality*, you will ask yourself: "How does he know what it feels like to be I_{131}?". $|I_{131}\rangle$ points at <u>any</u> state of consciousness of being unambiguously aware that the atom is I_{131} and not Xe_{131}. <u>It points at consciousness because there is no</u>

other reality. And conversely for $|Xe_{131}>$. You could put a lump of sugar on your tongue and say: "That's sugar". Dogs can smell a single molecule. Pretend you could put one atom on your tongue and say "Yep, that's I_{131} all right. Nope, it just changed to Xe_{131}." More realistically, we would need an MRI machine that can see individual atoms. It pings the atom with radio waves and looks at the frequency of the echo which comes from the nucleus. You see a guy at the beach with headphones on and a wand looking for coins buried in the sand. Medical MRIs are much more complicated. It would display $\mathbf{I_{131}}$ on your computer screen and $\mathbf{Xe_{131}}$ when it decays. Doing this has no effect on what the quarks and gluons are doing in the nucleus. The MRI looks at the precession of the whole nucleus magnetic moment in the magnetic field of the machine – it doesn't mess with the quarks. The puny radio frequency photons have as much effect on the quarks in the nucleus as would a feather would have tickling a rhinoceros.

In quantum math, $|I_{131}>$ and $|Xe_{131}>$ are a complete set of eigenvectors of a 2x2 Hermitian matrix. α and β in the above linear superposition equation are amplitudes: they are not measures of reality (consciousness) but related to densities of awareness. If I_{131} had another decay mode, the eigenvectors would be different and there would be three of them, etc. We are assuming there are only two in this case. If there is a decay mode we are missing, our calculated amplitudes will be wrong. If we assume there is another decay mode and there isn't, they will also be wrong. If our MRI tells us there the atom is I_{131}, we can start a timer at 0 and set $\alpha = 1$ and $\beta = 0$. We will ignore phase here and assume α and β are real numbers. The half life of I_{131} being 8 days means when our timer reads t = 8, our computer will calculate $\alpha = 1/\sqrt{2}$ and $\beta = 1/\sqrt{2}$. So the Born ~~probability~~ weights will be $\alpha^2 = 1/2$ and $\beta^2 = 1/2$ at t = 8. The word *probability* only takes on meaning if we do the same experiment many times over, always starting the timer t=0, and do statistics on the results. We can use the same I_{131} atom from the last experiment for the next experiment. Its reality is outside space and time. It is not "in the physical world" as von Neumann would have you believe, getting older and older. If in the last experiment you observed Xe_{131} you will have to get another I_{131} somewhere. If you can find one left over from a supernova explosion 5 billion years ago, use it. It will be just as "new" as one the nucleosynthesis lab made for you a few minutes ago. If the observed Born probabilities are $\alpha^2 = 0.5$ and $\beta^2 = 0.5$, after 8 days, our theory is good.

This is where physical world metaphysics and physics part ways.

Opening your box and looking at you atom has absolutely no effect on it.

- Every time you look you, *tat tvam asi*, will consciously experience the atom being I_{131}
- Every time you look you, *tat tvam asi*, will consciously experience the atom being Xe_{131}

These states of consciousness do not bind in a single state of awareness because their pointers in Hilbert space are orthogonal. Opening the box and looking in it has no effect on the atom you are looking but it has an enormous effect on you. Each begins a new journey in awareness. If you see a Xe_{131} atom the cat will be dead. If you see an I_{131} atom the cat will be purring and not even hungry if you left enough food an water for 8 days. What does bind in awareness is what you remember happening the last time you opened the box and looked. If it was Xe_{131}, it will never be I_{131} again, so the story ends in boredom. This does not mean histories starting with I_{131} are not real: The conscious experience of them does not bind with your present or future conscious awareness.

A detailed quantum mechanical analysis of the nucleus may show that amplitude for to I_{131} decompose is a complicated function of the time and is zero in some time windows for picoseconds at a time. pico = very, very, very small. And the calculations are very, very, very difficult so we make an approximation that the amplitude for I_{131} is a continuum that decays exponentially forever. This makes the conscious experience of observing I_{131} to be a continuum having a Born ~~probability~~ weight that falls by a factor of 2 every 8 days forever, always starting with 1 at time = 0. Unlike physical objects, every I_{131} atom is brand new, no matter how old it is. No, it is not a physical object. No more is Schrödinger's cat.

Evolution and the Uncertainty Principle

These words of Bertrand Russell in 1903 predate the quantum mechanics:

That Man is the product of causes that had no prevision of the end they were achieving; that his origin, his growth, his hopes and fears, his loves and his beliefs, are but the outcome of accidental collocations of atoms; that no fire, no heroism, no intensity of thought and feeling, can preserve individual life beyond the grave; that all the labors of the ages, all the devotion, all the inspiration, all the noonday brightness of human genius, are destined to extinction in the vast death of the solar system, and that the whole temple of Man's achievement must inevitably be buried beneath the debris of a universe in ruins – all these things, if not quite beyond dispute, are yet so nearly certain that no philosophy which rejects them can hope to stand. Only within the scaffolding of these truths, only on the firm foundation of unyielding despair, can the soul's habitation henceforth be safely built.

Russell's view is grounded on belief in the ultimate reality of the physical world governed by ultimately chaotic laws. They stand on the foundation ... *his origin, his growth, his hopes and fears, his loves and his beliefs, are but the outcome of <u>accidental</u> collocations of atoms.* (My underline). This premise was formally underwritten in the quantum mechanics by Heisenberg with the Copenhagen Interpretation, by Dirac and von Neumann, and by Feynman, that quantum events in the light of von Neumann's Process I that the wave function of an observed system collapses unpredictably into an eigenvector of the observation operator. Unless the eigenstate is the only one possible, which one it collapses into is absolutely unpredictable. Thus, collocations of atoms, of which we are an example, evolve in a way that is fundamentally accidental and Russell is correct if Copenhagen is science.

It is not science, but metaphysics.

In 1957 Everett showed that von Neumann's Process I was redundant. Reality can be understood as governed by the linear Process II alone that is deterministic for all reality in all space and time. The words *outcome of accidental collocations of atoms* are meaningless in Process II. There is no accident and no chaos in Process II. Process I is the thread of states of awareness in the image of which Heisenberg, Dirac and von Neumann created reality. In this epistle we identify reality as consciousness as being described by Process II of von Neumann, which was given to us by Schrödinger.

. . . Man is the product of causes that had no prevision of the end they were achieving has no meaning in Process II. All that is real is the deterministic and inevitable outcome of the Big Bang and physical law. Prevision of any end achieved is in Process II although we always find ourselves at some point of a thread of Process I. We do not have the theoretical wisdom to infer all the steps Process II took in creating us, but we can look in the fossil record for the traces it left. What we find vindicates Darwin. There is nothing in Geneses that supports the belief that the human race descended from Adam and Eve. Cain's wife was not in the line of descendancy from Adam and Eve as were others in the land of Nod on the east of Eden and elsewhere as the fossil and historical record shows.

Religious belief is not truth; it is only a theory about God; it is not equivalent to faith in God: Such belief commonly, but not universally, reveals appalling ignorance of science. Proclaiming: "I believe in God, therefore what I believe is what God believes." is the hallmark of Satan. For those who are not seekers of truth because they believe they already know it, their belief can be rubbish in the image of which they create God's truth. Those who read the Bible would do well to do so in an honest attempt to learn the truth rather than look in it for confirmation of what they already believe, instilled in them by parental and clerical authority. As a child I was taught in grade schools that people in the Bible once lived for many centuries like six or eight hundred years. This happened in government schools in a country with no constitutional constraint against the government establishing religious belief. In the early history of the Bible people were largely in nomadic tribes and used lunar civil calendars because of the ease of determining the beginning of the month from new moon. Months survive in our present calendars and the tradition of determining the first day of the month still is observed in Islam. Determining when a year has elapsed is more difficult but important because of synchronicity with the seasons which impact agriculture. This required a court astronomer with an observatory to house instruments such as astrolabes and hourglasses and records protected by the court police from vandalism. Nomads could not do this. The alternative was a vandal proof astrolabe like Stonehenge. In the later Biblical books, when calendars based on the year were available, people lives were numbered in years and not synodic months. The recidivists bungled the units, but hey, so did NASA when they botched a satellite launch by confusing English and metric units.

However much those who wrote the Bible or Koran were inspired by faith in God, every word was written by a pen held in a human hand controlled by a human brain. However conscientiously these words were written, their minds had no models of experience beyond myth, history and genealogy. The pursuit of truth over six thousand years has brought us to an understanding that would have been incomprehensible to them.

To uphold scripture as God's Absolute Truth contravenes the commandment:

> You shall not make for yourself any graven image,
>
> or any likeness of anything that is in heaven above,
>
> or that is in the earth beneath,
>
> or that is in the water under the earth; . . . . .

Like the worship of the one's interpretation of scientific theory as Absolute Truth, the worship of the myths of thousands of years ago as

Absolute Truth is an act of idolatry.

As a teenager I was inspired by the lucidity and simplicity of the words of Bertrand Russell. When he wrote them the quantum mechanics was still in the womb. When I read them I did not understand the quantum mechanics. As I read them now, they seem to be a view of reality from a very distant and different past. I have rewritten them as I think he may have written them now:

That we are the product of causes for which we had no prevision of the end achieved; that our origin, our growth, our hopes and fears, our loves and our beliefs, are the outcome of apodictic laws of consciousness; that no fire, no heroism, no intensity of thought and feeling, is needed to preserve consciousness beyond the grave; that the labors of the ages, all the devotion, all the inspiration, all the noonday brightness of human genius, has a path to destiny in conscious immortality that is eternally real, not inside the time of the birth and death of the solar system nor the birth and death of the physical universe, which has no ultimate reality – all these things, if not quite beyond dispute, are yet so nearly certain that no philosophy which rejects them can hope to stand. Only within the scaffolding of these truths, only on the firm foundation of unyielding hope, can the soul's habitation henceforth be safely built.

Conclusion

The Physical World

Dr. Johnson kicked a rock and took the rebound of his foot to be proof of the ultimate reality of the physical world. Why did he not take the conscious experience of his rebounding foot to be proof of the reality of consciousness, without which he would have experienced nothing? He would have replied that yes, the conscious experience happened, but is was caused by what is ultimately real: Physical Reality. Einstein and Newton would have agreed with him. By the end of the 20th century the classical physical world was dead. Johnson's rock, and every other rock, are made of the same stuff as consciousness: Absolute Nothingness. No! Wait: The quantum physical world may not be dead. There are such things as electrons, photons, quarks and gluons, you know. Take a proton or neutron: Three quarks bound together by gluons. A busy little thing, but I will never know what it feels like to be one. Or even know what it feels like to be you. But there is a configuration of electrons, photons, quarks and gluons that experiences consciousness with such certainty that belief about it is meaningless. I am it. I do believe, and I am pretty sure I am right but have no proof, it is also you. If you can read this I am certain you are not IBM, Inc; Du Pont, Inc; Wells Fargo Bank; that you are human.

How did God create the Physical Universe?

The quantum mechanics tell us: By making eigenstates orthogonal. This results in awareness that we call the physical world, in which nothing, past, present or future, is inconsistent. That is what God did. We did the rest: Create reality is its image. That is not a reality God created. It is a delusion. What God did create is consciousness: That is reality. It is a miracle. The biggest step in creating the physical universe was the creation of human awareness: took 5,000,000,000 years including making the atoms we are made of.

This is how <u>we</u> created the physical universe: First, lets get rid of infinities which are no problem for God but cause people problems. We define a new constant of nature: The temposity of light = τ. It is the time it takes light to travel unit distance, the reciprocal of the velocity of light. The next step in the creation of our central nervous systems was to approximate three SI unit constants:

- $\tau = 0.0000000033$ = temposity of light
- $h = 0.000\ 000\ 000\ 000\ 000\ 000\ 000\ 000\ 000\ 000\ 000\ 66$ = Planck's constant

- $\kappa = 0.000157$ = curvature of earth. Moon, mars a bit bigger.

with the approximation $\tau=h=\kappa=0$.

As soon as this accurate approximation is made: Voilà: The intuitive physical world jumps out of absolute nothingness into awareness! God is not lying to us: It is never manifest except in consciousness. But we look right through consciousness at what be believe is ultimate reality.

When we set these constants to their correct values, the intellectual physical world vanishes back into absolute nothingness. We have it on the authority of A. Einstein, that its comprehensibility is a miracle. This leaves unanswered the question of whether consciousness realizes absolute nothingness or absolute nothingness realizes consciousness. I asked my Zen master to show me the path to the answer. He said: "Finish your supper first. After that, lick your spoon".

A photon having passed a beam splitter may be said to exist in two superposed states, say $|R>$ and $|T>$, representing reflected and transmitted. In this book, we associate each state with the awareness of being in that state. As far as imagination goes, they self imagine the photon. We may think of each electron and each quark in Johnson's rock as imagining itself, and the common wave function imagining the rock. It imagines nothing beyond that. Now take a baboon that weighs as much as the rock: His atoms not only imagine themselves, but they imagine a heckuva lot more of the world beyond himself, especially that leopard over there. He is not indulging in his conscious experience of seeing the leopard: How pretty are her spots. He believes her to be a physical object in the physical word that presents an imminent danger to him, also a physical object. His self imagination is acute and accurate, using a theory of awareness that makes accurate predictions using three approximations that are utterly irrelevant to baboons: $\tau=h=\kappa=0$. The baboon's conscious experience of the leopard is absolutely real, but right now, not being aware that he is conscious is the least of his concerns.

The Physical World is created in the image of this theory: In reality, there is no such place, but believing otherwise is excellent baboon physics. To remind the baboon that Divine Providence or evolution, as you may prefer, has not forgotten to give him conscious awareness, give him a banana.

Beyond physics and science, moving the basis of reality from the physical world to consciousness has profound consequence which I cannot discuss at length, nor claim that I fully understand. Because of the enormous diversity of life on earth and the abundance of already discovered exoplanets, it is very improbable that life more advanced than ours does not abound in the universe. Sorry about the double negative, but I want to point at the improbability of being alone. Where such life does exist, the simplicity of what we call Newtonian physics will be known. They will continue to use this physics in most applied mathematics. The inevitable investigation of atomic spectra will lead to the quantum mechanics as it did for us. Understanding the distinction between consciousness and awareness as defined here will lead to throwing off the shackles of "physical reality", transcending to reality in the spiritual realm. This will be construed by some on earth as the fulfillment of Biblical prophesy of the end of the world. I do not know what the Biblical *end of the world* means. The conclusion reached here is forced on us by quantum mathematics because it represents conscious experience.

Things get worse for the *physical universe* when we look at the Special Theory. If there are three stars each 100 light years from one another, and we live on a planet of one of them, it is meaningless to talk about what is happening **now** on the other two stars, which our physical world model assumes we can, unless we are talking that are true of all possible worlds, like eclipses. We cannot even decide on the temporal ordering of events that happen on the other two stars so causality between these events is meaningless. We have to allow 100 years to elapse before we can enter into a physical world model of our common experience. Yet we go on talking about the physical world when look at the cosmos. We cannot enter into a physical world model that includes the quasars with anything that has happened to them in 13,000,000,000 years. When we recognize that all paths in consciousness are real, the *physical world* evaporates into nonsense.

The Physical World is a model of reality that makes accurate predictions of ordinary experience in the local neighborhood. It is consistent with Euclidean Geometry, Galilean kinematics, Newtonian mechanics, Newtonian gravitation and Maxwellian electromagnetic dynamics. At a mathematical formalism it exits: Only reality created in its image is delusion. It makes accurate models of electron trajectories in electron microscope optics, the dynamics of gas molecules, aircraft design, the motion of the planets and binary stars, the rotation of galaxies and the motion of galaxies in galactic clusters under dark matter gravitational force. It is a simplifying approximation to the quantum mechanics and the theories of relativity, made by setting $\tau=h=\kappa=0$, creating the Physical World which accurately predicts what will happen in consciousness where those approximations are justified. If we keep h = 0, but set τ = 1 in natural units, we still are in a sort of physical world where A is bigger than B and B is bigger than A. This is the real Wonderland of Alice we can see as absolutely real if we create reality in its image. Now set h = 1 and we are in a physical world that is no Wonderland, but a hell hole of random, superluminal causality: We must keep a stiff upper lip fighting armies of metaphysical myrmidons: demonic soldiers of Copenhagen that fight for darkness against enlightenment. Or stop creating reality in the image of Physical World $\tau=h=\kappa=0$ bunkum. You do not have to create reality in the image of anything: Look around, listen, smell the roses: That is reality and you didn't have to create it.

It would be an act of stupidity to abandon the physical world where we know it works: Quantum mechanics and relativity mathematics would give no meaningful increase in accuracy in the above cases but would incur enormous increase in computational expense. We never will abandon it. If there is intelligent life on other worlds more advanced than us, we may be sure they use it. The events of the past century have shown us that it is wrong. In some domains of experience it is worse even than a bad approximation. In predicting incandescent radiation it was a catastrophe. Contractors and architects routinely use the $\kappa=0$ approximation that the plumb lines are parallel but none (one would hope) believe the earth is flat. It is time for the rest of us to give up believing in the ultimate reality of the physical world.

The most compelling reason not to abandon the Physical World as a working model of experience is that it is hardwired into our central nervous system and manifest in intuition. Intuitively we feel the earth is flat. The intelligence that created us was not ours; it did not ask for our advice: had It been stupid enough to ask us for our advice, we would not be here; It did not make a mistake. We were never lied to: The Physical World is never manifest to us except in consciousness: We have the conscious experience of being in it. The only mistake we make is to create God in its image. We may believe that God forgives our idolatry, or God does not care, or God isn't there, but you should not forgive yourself for doing so because the proof of reality of consciousness is given only to you. Only you can seek its origin. You cannot give it to anyone else. What you find is your business. It is none of mine or anyone's else. Ergo it is none of the business of the government, which in the United States of America is proclaimed in the First Amendment. If *Tat tvam asi.* says the **conscious** experience of heaven and hell are equally real, the important thing is to be **aware** that you are in heaven or at least aware you are not in hell. Do not ask me how. I didn't write that book. And I have a bad track record keeping out of hell.

Evolution and Creationism

The subjective content of consciousness of animate beings is attuned to their survival: The amorous and erotic for procreation, the gustatory for sustenance, the fraternal for union, the visual for navigation, the martial for defense, the auditory for communication and so on. We sense the physical only to the extent that it is necessary for basic need. Conscious beings see and feel themselves to be in a three dimensional world containing ponderable objects with absolute attributes of mass, location in space and movement in time. Touch, smell, taste, sensation of temperature tell us what we need to know about the things we come in contact with. These conscious states were imbued in us by evolution, or by God if you wish. They are not scientifically observable. Although it does not always have to be so, present day scientific understanding of the relation of the subjective content of consciousness to evolution is non-existent. Plants exhibit awareness, although not on the short time scales of neural activity, so a belief that the subjective content of consciousness played no role in their evolution is not science but a wild guess.

A century before I was born, it was believed, even by learned scientists, that living things were governed by a Vital Force: divine laws incomprehensible to mortals. Then Friedrich Wöhler synthesized urea from inorganic components. In the hundred years from the mid 19th to the mid 20th century, the greatest discovery in the history of science unfolded. Like the quantum mechanics, it took genius to wring understanding out of empirical data. Unlike the quantum mechanics it did not lead to a grand theory, but something more astonishing: The understanding of the role played by every single H, C, N and O atom in structure of the genetic code and the role the codons played in the synthesis of proteins from amino acids, also mainly H, C, N and O atoms, in the construction of every living thing on earth. The quantum mechanics supplied understanding of the configuration and binding of the atoms of nucleotide bases and amino acids. It gives cause for profound respect for Einstein's insight that the comprehensibility of reality is a miracle.

The quantum chemistry of proteins challenges modeling with foreseeable computers, but we can observe what happens in living things. To make models tractable we can use stochastic formulations and statistic analysis when our abilities fail to model reality exactly, and the amount of data is overwhelming. We can throw around words like chaotic, accidental, random, uncertain with justification because they describe the <u>theories</u> that we use to represent data. They do not describe <u>reality</u>. Then we can do something we are very good at: Create reality in the image of our theories about it, and uphold as divine truth Principles of Uncertainty; we are Accidental Configurations of atoms.

We owe Everett a debt of gratitude for showing us that the mathematical underpinning of the quantum mechanics is deterministic. There is no room in it for uncertainty, accident, randomness. It now shows us that every observation we make embarks us on a transcendent infinitude of journeys in consciousness caused by eigenstates of the observation we make, with amplitudes determined by the thing with which we interact. That all these journeys into the future are equally real. This is an understanding of reality that beggars the physical world into insignificance. We can look back along the single path that has brought us over 14,000,000,000 years from the Big Bang to this planet of a G2V star of a barred spiral galaxy. We can remember nothing of the future. Q. Why are we here now?   A. Because it is real.   But so is every other path in consciousness with which our awareness does not bind. Ours is the single terminal of a deterministic journey that put the enormous diversity of life on earth; the departure point of infinitely many journeys into the future: Consciousness is deterministic; in awareness we are free to make choices. There is every reason to believe that every planet in the universe with conditions as favorable as earth will have the same diversity of life.

Before people of science pity those who believe that Eve was made out of Adam's rib 6000 years ago and they got married and we are their descendants (don't even think about Cain's wife), in other words Creationism, we need to own to the equally mind stifling nonsense we believe: The Copenhagen Interpretation and its child: Our random walk of accidentally configured atoms, our every step eradicating all the other steps the quantum mechanics says are real, lest their reality kill the god we worship: The Physical World. Father Feynman is watching you lest you ask "How can it be like that?" Dare ask, and you will be sent down his drain. But, if you ever do go down there, please look for him and pull him out.

Why consciousness now?

A legitimate question is: If consciousness is ultimate reality, why did the founding fathers of modern physics not come to that conclusion long ago? Von Neumann almost did. Here are the views they did come to, seen in the light of the definitions of awareness and consciousness given in this epistle:

Long ago Wigner gave the view that the collapse of the wave function did not occur until awareness entered human consciousness.

Heisenberg saw quantum mechanics as modeling awareness as the ultimate basis of experience, forcing him to the Copenhagen Interpretation.

Born's probabilistic interpretation of predicted amplitudes, mathematically correct as it applies to statistics of a sequences of observed data in awareness, did nothing to illuminate the role of consciousness.

Schrödinger wrote about consciousness and gave us the most perceptive analysis of consciousness and conscious self-hood as it relates to science of any of the quantum pioneers, if not scientists. However, that I am aware of, he did not tie it into the quantum mechanics. What he wrote on consciousness is hard to find. I wish that someone with the ability to do so would publish a complete collection of these writings.

In the sense in which I have defined the word consciousness, Einstein gave no thought to it as it is related to physics. Where he used the word consciousness in other context, one cannot be sure that he was not talking about awareness as I have defined it. He considered reality to be physical; definitely not spiritual. Yet he spoke of the miraculous. He saw the Copenhagen Interpretation as flying in the face of physical reality and sought escape from its conclusions by seeing quantum mechanics as an incomplete theory requiring additional parameters to explain observations. He was never aware that doing so made predictions which, after his time, were to be contradicted by observational evidence. In retrospect we can see that his error was to try to drag the quantum mechanics back into classical science, failing to understand that the quantum experience was teaching us a new truth: a new way of framing science that classical science could not represent.

The only grand master of the quantum game who gave consciousness serious and conscientious consideration was not a physicist, but the mathematician von Neumann. However he placed consciousness into physical reality reinforcing Dirac's conclusions that causality is broken and reality is intrinsically random. I have used the word *random* to avoid words like *stochastic* and *probabilistic* that may imply lack of sufficient information about something certain. The Uncertainty Principle implies uncertainty so profound that we do not, at least in English, have a word to describe it. One has to invoke God by saying only God can know, or even God cannot know. This uncertainty is metaphysical delusion.

Everett's additions to quantum math did not come for another 25 years. He showed that the collapse of the system state vector is unneeded metaphysics. He supported von Neumann's respect for the principle of psycho-physical parallelism but gave no further consideration to consciousness in his lifetime that I am aware of beyond recognizing the subjective reality of awareness caused by each eigenstate of the observed system.

The Continuum

If every quantum event creates infinitely many journeys in consciousness, where is God going to put all of them? Hilbert, of Hilbert Space fame, has the answer: "In my hotel which has infinitely many rooms". When the hotel is full, and a new guest arrives, Hilbert moves every guest into the next higher numbered room. This frees up room 1 for the new guest. If infinitely many new guests arrive, Hilbert asks present guests to move to a room number twice their present number. This frees up all the odd numbered rooms of which there are infinitely many for the new guests. This can go on forever. Worrying about where God is going to put all that consciousness is like worrying about where God is going to put all those numbers. Consciousness and numbers are made of nothingness. When you go to heaven, you will hear God singing "Oh, I got plenty o' nuttin' And nuttin's plenty for me".

The ancients believed that all numbers, integers and fractions, could be represented by the ratio of two integers like 3/2, 5/1, 22/7, 553/311, called rational numbers. Then they discovered a proof that the square root of two cannot be so represented. It is not complicated – you can find it online: ***Prove square root of 2 irrational.*** Such numbers are called irrational and have infinitely many digits with no repeating patterns. It was shown by Cantor and others in the 1800s that the irrationals are uncountably more infinite than the rationals. They form a dense continuum. This is the continuum of the position of a free particle in the quantum mechanics in which each value of a coordinate is an eigenstate orthogonal to all others. We have no choice to accept as real (in the reality sense) each eigenstate for which the wave-function is non zero. We identify this reality as consciousness which forms a continuum over all eigenstates. When we observe the same system in the same way, the same thing always happens: We consciously experience being in each eigenstate, but each conscious state does not bind in awareness with any other. We thus are aware of the particle having a discrete eigenvalue of the continuum and are aware that it is no other, yet all are equally real. This is the interpretation Everett gave us if we see reality as consciousness. There is absolutely no resemblance between this reality of spirit and the Physical World. We cannot even create the Physical World in the image of quantum awareness unless we invoke our two friendly demigods enforcing the Anthropic Principle: Uncertaintella to randomly select some eigenstate to be the only one that remains real. **Hey Tella! That's always me!** Until she does she cannot allow premature collapse of the state vector even by God's premonition. Once she does, she calls on her monster Collapzilla to gobble up instantly all the other eigenstates she rejected as candidates for reality. **Hey, Zilla! Them, not me.** This is the hocuspocus we must buy into to believe in the ultimate reality of the Physical World. As a model that represents experience, we can claim that it exists. In the sense that it does, the spirit world does not. The world of spirit is real. The Physical World has no reality beyond a mathematical approximation.

We have to distinguish consciousness and awareness. Believing the earth is flat is an absolutely real conscious experience. Being aware that it is flat is an OK theory for architects, but hölynpöly for Finnish aströnömers.

In the beginning

Those who look for an accounting of consciousness cannot be reminded too often of Schrödinger's words:

Consciousness cannot be accounted for in physical terms
For consciousness is absolutely fundamental
It cannot be accounted for in terms of anything else

In the beginning absolutely nothing existed. Call it the Void. This view, once the domain of mysticism, has gained increasing currency in physics over the past forty years. Most physicists today believe that the Void was disrupted quite suddenly, 14,000,000,000 years ago: There was a Big Bang and the ultimately real Physical Universe, future home of the steam locomotive, sprang out of the Void, became real and has been around ever since. This view that the Physical Universe is ultimate reality is not supported by the arguments of this paper. The Big Bang is seen as the origin of all paths in consciousness. The existential structure of consciousness, as best we understand it, is the quantum mechanics which describes the interaction of the elementary particles seen as mathematically existential properties of this structure. Their relationship to the Big Bang is beyond my competence and is not relevant to this book. Those interested in this question I refer to the ideas of others, an example of which is a theory of Hans Dehmelt: "The world-atom, a tightly bound cosmon/anticosmon pair of zero relativistic total mass, arose from the nothing state in a quantum jump. Rapid decay of the pair launched the big bang and created the universe." I disqualify myself from forming an opinion of such theory.

The following view is what is left by Occam's Razor notwithstanding the Big Bang and the fact that the steam locomotive, having no need of the quantum mechanics, can blow it away with a puff of steam: <u>Not only is it true that, in the beginning absolutely nothing existed, but in reality absolutely nothing ever has existed and absolutely nothing ever will</u>, if we understand *existence* to be the ***physical world.*** If we understand by existence what mathematicians mean by that word, we do not need the Big Bang to be its origin. We are the apodictic consequence of the Void. The Void and not the Physical World is the homeland of truth accessible to us as the properties of numbers and their handmaidens the elementary particles which are mathematically exact entities. The Void is the substance of consciousness which, although absolutely real, is made of Absolute Nothingness. Consciousness is not some existential substance like ectoplasm in a physical world. The reality of consciousness is a miracle upon which atoms of the physical order dance attendance. Their dance is the structure of reality.

The bridge has never been built between the properties of numbers and the exact mathematical properties of the elementary particles of the physical order: leptons, quarks and bosons. String theorists have never given us a theory that is in demonstrable correspondence with what we observe as did classical physics and the quantum mechanics and the theories of relativity. However, if such a bridge can be built, string theorists as bridge builders are our best hope. It would be the theory of everything. But it would not explain the reality of consciousness, which always will be miraculous. It will never be otherwise because we cannot observe it.

Belief and Truth

One is not crazy to believe something we know is not true, if knowing how it could be true is beyond our comprehension. When quasars were discovered, astronomers divided 50/50 into two vehemently opinionated strongly disagreeing groups. Obviously both could not have been right and both could have been wrong. Long ago Sherrington pointed out that this always happens in science when there is no rational basis to choose between the alternatives. This is very common in elections which often split very close to 50/50 for the same reason, notwithstanding the vehement extremes of political opinion of voters. In science, belief is the motivation to do experiments that reveal truth. One is not going to be very motivated to prove oneself wrong, so we can define belief in science as the motivation to design experiments. In science, belief ultimately defers to the theory that represents the experimental evidence, but it does not die easily as the experience of the quantum mechanics has shown. Politics is another story.

We are free to believe truth, fantasy, rubbish; to believe pernicious falsehood.

We are not free to ordain the truth.

We are free to ask any question.

The answer depends on the question asked.

Once the question is asked, as Omar Khayyám told us a thousand years ago, we are not free, to ordain the answer.

The Moving Finger writes; and, having writ,
Moves on: nor all thy Piety nor Wit
Shall lure it back to cancel half a Line,
Nor all thy Tears wash out a Word of it.

Q: When did reality begin and when will it end?   A: It never began and it never will end. How can Absolute Nothingness ever begin? How can it ever end?

However, even Absolute Nothingness is a crutch that we limp on after they took the physical world away from us. Throw away the crutch. The reality of consciousness needs no crutch.

Looking to our future, all quantum mechanical paths from where we are are open to us. The quantum mechanics predicts the reality of all other paths in consciousness as certainly as it predicts ours. Your seeking will chart your course and the moving finger will write your destiny. If there is a path you do not want to follow, do not ask the question that opens its door. It is an illusion to believe that whereas the present moment of consciousness is real but a moment ago it was not and a moment hence it will cease to be. All moments of consciousness are equally real from the beginning to the and of time in all possible worlds. Consciousness is not inside space and time, but awareness is. The journeys we make in space and time are journeys in awareness. We are not arbitrators of the structure of consciousness, but we have the choice not to overtake on a blind curve if we do not want the moving finger to write our obituary.

What is ultimately important to us are the choices we make in awareness. We need not be worried about consciousness because, no matter what happens, it will always be real. Saying "it does not matter to me what happens after I am dead" is a cop out. You cannot escape from consciousness. There is Absolutely Nothing to escape from.

Ponder this koan:

It does not matter what happens, it will happen to you.

It does matter what happens, because it will happen to you.

Parable of Physical Reality

In the jungle there is a tree with two branches each of which has two branches and so on for twenty levels of branching, ending in a million leaves. On each leaf live bugs that believe theirs is the only leaf. The came up the trunk and remember whether they turned left or right at every branch which sticks in their brains as a twenty bit binary number which they worship as scripture. They believe it is the only number selected by God that leads to Reality.

The dreams of reality and the reality of dreams

We should not allow our dreams to keep us awake, but leave them in dreamland where, before our minds, we live in the Physical Universe which we worship as our Creator and the Foundation of our being.

The quantum mechanics is our wake up call: To awaken into a dream before the mind of God: That we are made of the stuff dreams are made of, as the bard told us long ago. It is a dream more real than any physical world could be.

God

Whenever we try to understand anything, we are trying to understand what God has wrought. We cannot without presumption stand outside God to do this. It may appear that this epistle is an invocation to believe in God. I have learned from my own bitter experience, that belief in God commonly means the creation, or destruction, of God in the image of ones beliefs. Since childhood and to this day I listen to politicians clothe their despicable beliefs with God's name. They defiled the newly created United States of America, requiring the First Amendment of the Constitution to stop this abuse of power. I write this in the November of the ISIL atrocities in Paris when psychotic criminals machine gunned crowds while chanting *Allah Akbar.* Their barbaric self-righteousness inspires me never to proclaim that I believe in God, even if I do. Such proclamation never glorifies God, but only glorifies oneself. Max Born said: *The belief that there is only one truth, and that oneself is in possession of it, is the root of all evil in the world.* Whatever they believe about God, if God does not believe in them, it does not matter what they believe.

However, I will proclaim that there is no other source of truth. Faith in God is beyond question: It can only be judged by God. God's truth does not come out of gun barrel or the mouth of a lunatic. The destiny of life in the universe is truth's pursuit. This journey ends in the belief that we already know it: It is a dead end road both in religion and science and currently a dead end road in physics. The purpose of this book is to cast light on ideas that I have found fruitful in understanding consciousness; its purpose is not to instill belief in your mind. What you choose to believe is none of my business.

Jesus said: "Ask, and it will be given to you; seek, and you will find; knock, and it will be opened to you. For everyone who asks receives, and he who seeks finds, and to him who knocks it will be opened....". The history of science stands as a monument to the truth of these words. The results of scientific experiment do not choose between good and evil. They empower both; that choice is the responsibility of the experimenter, not the experiment. There are no bounds on what we can comprehend. Einstein told us: It is miraculous that we can comprehend anything. Beyond anything we can comprehend there lies an infinite realm of eternal magic that is absolute truth.

Appendix A: Technical terms

Angular momentum – See momentum

Canonical – Pairs of first order equations in Hamilton's version of classical mechanics which replace one second order differential equation in Newton's version. A pair relates a coordinate and its *conjugate* momentum.

Capacitor – two electrically conductive plates separated by an electrical insulator

Classical kinematics – If two frames of reference are in relative motion at vector velocity v, points in one frame will move through the other by the vector distance d in time t by the amount v.t. Instruments measuring distance and time in each frame will synchronize

Coefficient - A constant that multiplies a variable giving some other variable equal to the coefficient when the first variable is 1

Conjugate– the relationship between a coordinate and its conjugate momentum in classical physics – they both must be known at the same time to determine the system motion. This is impossible in the quantum mechanics: See *complementarity*

Complex conjugate of an object – gotten by changing the sign of $i=\sqrt{-1}$ where it appears everywhere in the object. The object may be as simple as a complex number of the form $x+i.y$

Complementarity – the relationship between coordinates in the quantum mechanic equivalent to *conjugate* in classical mechanics. The operators for complimentary coordinates do not commute

Coordinates - Variable numbers like distances or angles that specify the changing state of a mechanical system as distinct from constants that specify the unchanging part of its architecture

Degree of freedom – a changeable aspect of a system that requires a coordinate to specify it

Differential equation – an equation describing the relationships of the coordinates of a system and their rates of change with respect to time and/or other coordinates. If F and G are solutions to the equation, it is called linear if F+G is also a solution.

Dynamics The way *kinematics* of massive objects is affected by the forces acting on them.

Eigenfunctions - A property of an operator: Operating on one of its eigenfunctions does not change it except by scaling by a constant called the *eigenvalue*

Eigenstate – The state of a system described by a pure eigenfunction as distinct from a linear superposition of them

Eigenvalues – see eigenfunction

Electric charge – a fundamental property of elementary particles. It is quantized in units of ...-2, -1, 0, 1, 2.... of the electron charge.

Like point charges repel and unlike attract, following an inverse square law of force. The charge content of the universe is zero. Total charge is conserved in interactions between particles

Electrostatic – an adjective describing a force arising from electric change

Energy – Classically: a scalar = $m.v^2/2$. It is conserved in elementary particle interactions. Each particle of mass m has an associated energy of $m.c^2$ where c is the velocity of light

Force – the time rate of change of momentum, or the distance rate of change of energy

Fourier transforms – Periodic functions of, say, time have been known since antiquity to be representable by series of wave functions of time but expressed as functions each with a different wave frequency. Fourier, circa 1800, showed how to determine the coefficients of the series. Non periodic functions can be so represented in which the frequencies are not discrete but a continuum, called a transform

Function – A relationship between numbers either given as a mathematical formula or a printed table

Galvanometer – A sensitive device for measuring the flow of electric change

Great circle – A circle on the surface of a sphere dividing the sphere into two hemispheres

Hermitian – A matrix which is the complex conjugate of its transpose (flip about the upper left to lower right diagonal). Its eigenvalues, which represent observations, are real numbers. All quantum mechanical measurement operators have this property

Hilbert Space - A space spanned by a complete set of eigenvectors characterized by the inner or scalar product.

Inner product – also the scalar or dot product of two vectors:

(x,y,z...) and (a, b, c...) = $a.x + b.y + cz$ ... = a *scalar*. See Appendix E

Interference – An elementary particle existing in multiple eigenstates all of which cause a single effect with an amplitude equal to the sum of the amplitudes for each eigenstate. This can cause cancellation or reinforcement of amplitude depending on where or when the effect is observed, called an interference pattern

Interferometer - A device to exhibit wave interference

Kinematics – The theory of time dependent geometry. Kinematic theory known since antiquity was corrected by Einstein in 1905.

Linear independence – A set of vectors are linearly independent if none can be represented as a linear superposition of the others

Linear superposition – if **x,y,z,...** are linearly independent vectors, then a**x**+b**y**+c**z**... is a linear superposition of them where a,b,c... are constants that characterize the superposition

Magnetostatic - an adjective describing a force arising from a fixed magnet

Matrix mechanics – The Heisenberg, Born, Jordan theory of the quantum mechanics

Momentum – *Classically:* A mass m moving with velocity v has vector momentum m.v. If the vector **v** is tangent to a circle of radius r about some point P, the angular momentum of m about P is m.v.r. In classical mechanics, both the position and its conjugate momentum must be known at the same time to uniquely determine future motion. Quantum mechanically this is impossible because generally there is no unique future motion. The operator for x direction momentum p is $\partial/\partial x$. The operator for x dimension position x is $\partial/\partial p$. These two operators do not commute. This is called the "Uncertainty" Principle. The concept that uncertainty is implied by the non-commutativity of complementary operators is metaphysics.

Operators – A mathematical device that operates on a function or vector to produce another function or vector

Orthogonal – two vectors are orthogonal if their inner product is zero. Spatial vectors that are orthogonal form a right angle to one another.

Photon – The elementary particle that mitigates the electromagnetic force.

Quantum amplitudes – every event has a (non relativistic) quantum mechanical amplitude, generally a complex number, which multiplied by its complex conjugate is the Born probability to happen.

Rate of change of y with respect to x – the ratio dy/dx where dx is an infinitesimal change in x that caused y to change by the infinitesimal amount dy. If there are other variables that can change y, the ratio is written $\partial y / \partial x$.

Real numbers – ordinary counting numbers and fractions that do not contain the square root of -1

Relativistic – The domain of mechanics and kinematics where speeds are a significant fraction of the speed of light requiring the Special Theory of Relativity. Relativistic kinematics changes the equations of *classical kinematics* to a form discovered by Lorentz, and called the Lorentz transformations, later derived from first principles by A. Einstein. See Appendix B.

State vector – a vector in *Hilbert Space* that is a complete description of the state of a quantum mechanical system

Scalar – a single number

Scalar product – see *inner product* of two vectors:

Solution – a function that satisfies a differential equation is its solution. Numbers that satisfy an algebraic equation are a solution. For example, $x^2 = 1$ has the solutions +1 and -1. $x^2 = -1$ has no real solution but does have the solutions +i and -i where i is the square root of -1.

Vector product – a form of product of two vectors which is another vector orthogonal to both

Vector - a list of numbers

Wave-function – same as state vector – a solution of Schrödinger's equation

Appendix B: Lorentz Transformations

A reference frame with three orthogonal coordinates x, y, z, t is in motion relative to another ξ, η, ζ, τ with the same axis orientation at velocity v in the z or ζ direction. Observers in both systems will agree that the relative velocity v is the same in both systems, and that the x, and y coordinates are the same in both systems. But they will disagree about the z, ζ coordinate and the time t, t.

$\xi = x$

$\eta = y$

$\zeta = \gamma.(z - v.t)$ Classical kinematics: $z = (z - vt)$

$\tau = \gamma.[t - (v/c).(z/c)]$ Classical kinematics: $t = t$

where c is the velocity of light:

$\gamma = 1/\sqrt{[1 - (v/c)^2]}$ If v is small, $\gamma \approx 1$, but goes to 0 as v approaches c.

Distance in the z direction is the same as classical kinematics but shrunk by a factor γ.

If we measure velocity in units of c and distance in units of time light takes to travel the distance, the time equation becomes:

$\tau = \gamma.[t - v.z]$

The humdinger is that the above equation is the same as the z equation, with z and t interchanging roles:

$\zeta = \gamma.(z - v.t)$

v.t is the distance we have traveled in the z direction and is negative because it reduces the distance to out destination. When v.t = z, we have arrived. What came as a surprise is that $\tau = \gamma.[t - v.z]$ shows we are also traveling through time of the other system. Classically, $\tau = t$. Time is the same in both systems.

It is an instructive exercise in high school algebra to transform these equations to give x,y,z,t in terms of ξ,η,ζ,τ. The resulting equations are the same which accords with the principle of relativity.

$x = \xi$

$y = \eta$

$z = \gamma.(\zeta - v.t)$ Classical kinematics: $z = (\zeta - v.\tau)$

$t = \gamma.[\tau - (v/c).(\zeta/c)]$ Classical kinematics: $t = \tau$

v is the same in both systems.

Appendix C: Infinitesimal Calculus

Functions

A function is a relationship between numbers either given as a mathematical formula or a printed table. A bus timetable is a function – the time of arrival of a bus as a function of the bus stop number. If you look at how the number y varies as you vary x, y is a function of x. It is written y = f(x). f, or any other letter you wish to use, is just the function name and is not a number multiplying x. If that were intended, it should be written f.x.

Often one just writes y = y(x) where x is called the independent variable and y the dependent. You have to figure out from context that y(x) in this case does not mean y multiplied by x. Obviously x is also a function of y. Put this way, where y and x interchange roles, is called the *inverse function*. In a printed table just interchange the column for x with the column for y. If it is a formula or equation, getting the inverse may not be as easy as you think.

The terms *dependent* and *independent* are unfortunate because it may imply a causal connection between them. Here is a counter example. In a certain town it may be true that people who buy raincoats never buy umbrellas and those who buy umbrellas never buy raincoats. But one may find looking at daily sales of raincoats c and daily sales of umbrellas u, that 5 umbrellas are sold for every 2 raincoats. So:

$$u = 2.5.c$$

is the functional relationship between sales figures even though my buying a raincoat does not force you to buy 2.5 umbrellas. Hey, its raining. It may be true that more people wear raincoats but umbrellas do not last as long. Also the sale of leaf-bags b may go up when u goes up because windy days more umbrellas are wrecked than on days leaf-bags are not needed.

We could write u = f(c,b) to show that umbrella sales depend on two independent variables c and b.

Rates of change of functions

The laws of physics are, for the most part, relationships between rates of change. One should be aware of the following notation. If du is understood to be an increment in u, and dc an increment in c, if

$$u = 2.5.c$$

then

$$du = 2.5 \cdot dc$$

or the rate of change of u when you change r is:

$$du/dc = 2.5$$

du/dc is described with this jargon: *"the derivative of u with respect to c"*. This is the notion of Leibniz, co-inventor of the infinitesimal calculus with Newton. du and dc are called infinitesimal increments called *differentials.* Infinitesimal, which means arbitrarily small but not zero, would not be needed in the above example. Calling the ratio du/dc a *derivative* is nasty word so some call it the *differential coefficient. Diffco* or something would have been better, but *derivative* is most commonly used.

Newton used a different notation for differentiation with respect to time. If **a**, say is position, a with a dot over it is **a** velocity and a double dot like ä is acceleration. I cannot show you with this text editor and an umlaut does not work with any letter. This notation is commonly used by engineers but you may run into it, like **d^2a/dt^2 = ä.**

Differentials would be needed in this case:

$$u = 2.5 \cdot c^2$$

when the derivative of u with respect to c, aka the rate of change of u with c, aka the slope of c, is:

$$du/dc = 5 \cdot c$$

This is easy to prove. Take a simpler case $u = c^2$. If we change u by du, c will change by dc.

u+du = (c+dc)^2 , or

du = (c+dc)^2 − c^2 because u = c^2

= c^2 + 2c·dc + dc^2 − c^2

=   + 2c·dc + dc^2

But we want du/dc, so divide the above equation on each side by dc

du/dc =   + 2c + dc

Now make dc infinitely small but not 0, giving

du/dc =   + 2c

If we had $u = c^n$, where n is in any whole number, negative, positive but not zero the derivative is du/dc = $n \cdot c^{n-1}$. If you remember the binomial expansion from high school algebra, you can derive this as above. It is true of any function that if it is multiplied by a constant the derivative will be multiplied by the same constant which is why $d(2.5 \cdot c^2)/dc = 5 \cdot c$

d/dc can be thought of as an operator operating on the function following it, like this:

$$d/dc[2.5.c^2] = 2.5 \cdot d/dc[c^2] = 5.c$$

It is called the *differential operator* and the operation it performs is called *differentiation* in the differential calculus. These big word mean nothing more than we are trying to get the rate of change of the thing we are operating on, u, when we vary c, in this case. This operator is linear which means that the derivative of a sum of functions is the sum of their derivatives. It follows that the derivative of a constant times a function is the constant times its derivatives. It is not true that the derivative of a product of two functions, say **w** and **v**, is the product of derivatives. If **f = w.v, df is** not **dw.dv** but it is **df = w.dv + v.dw**. This is taught in calculus courses.

If we were given 5.c and asked to find what u is, the inverse of *differentiation*, the operation is called *integration* and written

$$u = \int (5.c)dc = 2.5.c^2$$

Here the fancy S written $\int$ means sum the areas of all the little rectangular slabs 5c high and dc wide to get the total area under the curve.

We will have no need here to get into this operation which is best left to mathematicians who publish books of the integrals of thousands of functions for the benefit of mere mortals.

The simplicity of physics is that its basic equations are differential and easy to write. Solving them is another story: This is done by integration: This is where you have to pay you taxes and it is seldom easy. Solving the differential equations of the motion of the moon, even with major funding from maritime governments, took 300 years.

If you differentiate a derivative, that is perform the d/dx operation twice the result in Leibnitz notation is written:

$$d/dx(d/dx) = d^2/dx^2$$

The notation d^2 does not mean d squared – it just means differentiate twice. But the dx^2 really does mean dx squared. Differentiating y seven times in a row you will write as

$$d^7y/dx^7$$

More than one independent variable

We can apply the methods of the calculus to the case where there is more than one independent variable, like u = f(c,b). Differentials are written with a different letter: ∂ when there is more than one independent variable. $\partial u/\partial c = \partial f(c,b)/\partial c$ is called the partial derivative of f with respect to c because we only look at how f changes when we only change the c **part** of what f depends on. For this reason ∂ is the symbol for **partial** derivatives. Likewise $\partial u/\partial b = \partial f(c,b)/\partial b$ is called the partial derivative of f with respect to b because we only look at how f changes when we change the b part of what f depends on. $\partial u/\partial b$ will still depend on c, i.e. vary when we vary c, but it was computed by looking at how much u varied when we varied only b, but did not vary c.

It may seem that the partial derivative is useless because if u = f(c,b) we want to know what happens if you change **both** c and b at the same time. However, partial derivatives are easy to get because they follow the same rule of differentiation as functions of a single variable by just treating all the other independent variables as constants. Once we know them, it is easy to get the total differential of f when we change both c and b by using what is called the chain rule which is this:

$$df = (\partial f/\partial c).dc + (\partial f/\partial b).db$$

Suppose f also depended on the time t and c and b were also time dependent. Then:

$$df/dt = (\partial f/\partial t) + (\partial f/\partial c).(dc/dt) + (\partial f/\partial b).(db/dt)$$

Showing that the total rate of change of f with time t, is more that its partial change when we change t, treating c and b as constant, but also is affected by the way c and b change with time. This looks very complicated, but if you stare at these equations for a while they become obvious. The reason that the terms are simply added is because when the changes in variables are infinitesimal, cross talk between them vanishes. This was Newton and Leibniz big insight. If we square an infinitesimal say dc, then $(dc)^2$ is infinitely smaller than dc so we can ignore it. Even more true if we raise it to a power higher than 2.

If you differentiate y with respect to b and then with respect to c write it as

$$(\partial^2 f/\partial b \partial c)$$

The exponential function

Above we showed if $y = x^n$, where n is in any whole number, negative, positive or zero the derivative is $dy/dx = n.x^{n-1}$ Multiply in front by the exponent and reduce the exponent by 1.

If $y = x^4$, $dy/dx = 4.x^3$

If $y = x^{-4}$, $dy/dx = -4.x^{-5}$

if n = 0 the function is a constant and the derivative is zero. The derivative of 6 is zero because 6 does not change when you change x.

This being true, what do we have to differentiate to get 1/x? This is equivalent to asking the question what is $\int dx/x$ because we cannot get 1/x by differentiating any power of x. If it had been left to me to answer this question, it would still be unanswered. It was answered by the mathematician Napier who said:

$$d(\text{logarithm } x)/dx = 1/x.$$

Now $10^3 = 1000$, so 3 is the logarithm of 1000, to the base 10,

And $2^3 = 8$, so 3 is the logarithm of 8, to the base 2.

But Napier's logarithms were not to any base as mundane as 2 or 10, but to this base:

e = 2.7182818284590452353602874713526624977572470936995....

approximately. Called e, it a fundamental mathematical constant and the base of what are called natural logarithms and is the lover of:

π=3.1415926535897932384626433832795028841971693993751058 2..

approximately, another fundamental mathematical constant I hope you learned about in grade school. Probably discovered by a caveman who, like his equivalent politician, thought it was 3 or 22/7.

Logarithmic functions to a base n are written $\log_n(x)$ with the base as a subscript. Natural logarithmic functions to the base e are written ln(x). x here is any independent variable.

So the inverse of the ln() function is the exponential function $y = e^x$ because x is the natural log of y. It is fairly easy to show that the derivative of e^x is e^x. It is its own derivative. It is the exponential growth function like compound interest. It is the least wavy curved function known. From this property it is possible to derive the following power series expansion of e^x

The function:

$e^x = 1 + x/(1) + x^2/(1.2) + x^3/(1.2.3) + x^4/(1.2.3.4) + x^5/(1.2.3.4.5) + \ldots$ and so on forever.

You can see at once that differentiating the power series gives back the same series.

Each term is the derivative of the term to its right. The derivative of the sum of functions is the sum of the derivatives.

Appendix D: √−1 makes waves

In grade school we are taught that:

A <u>positive</u> number times itself is <u>positive</u>

A <u>negative</u> number times itself is also <u>positive</u>

That's unfair! Why does not something times itself ever equal a <u>negative</u> number. If I asked Mrs. Tibbit this question in the 5th grade she would probably have told me not to be impertinent.

Mathematicians were forced to admit that there was such a number: Any number times √−1. Doing this greatly simplified the theory of equations. For example

$x^2 - 1 = 0$ has the solution $x = +1$ and -1

$x^2 + 1 = 0$ has no solution. Why not?

Well it does if you count in units of a new number called **i** = √−1. Now $x^2 + 1 = 0$ has the solution $x = +\mathbf{i}$ and $-\mathbf{i}$.

Introducing i into mathematics solved many problems and opened new vistas. At first is was met with resistance: √+1 was called a *real* number and √−1 an *imaginary* number, but it is really just as real as a real number. <u>This is a pain in the neck in this book because I need to use the word *real* in the completely different philosophic context of reality.</u> Counting in units of √−1, called imaginary, is just as real, but please don't tell a mathematician I said so. This unfortunate terminology stuck and is still with us. Numbers which count both units are called *complex*. Another misnomer because they are not very complex and obey all the usual rules of algebra – grade school kids can easily be taught to add and multiply them. But it was an attempt to correct the error of calling them imaginary.

It is true that

(+1).(+1) = +(1²) = (−1).(−1) = −(i²) = +1
(+i).(+i) = +(i²) = (−i).(−i) = −(1²) = −1

If you replace 1 with i and i with 1 in the parenthesis of the above equations you get:

(+i).(+i) = +(i²) = (−i).(−i) = −(1²) = −1
(+1).(+1) = +(1²) = (−1).(−1) = −(i²) = +1

which are exactly the same as above (only with order reversed). Thus it does not really matter whether you count, say, people, inches, kilowatts or onions in units of √+1 or we count them in units of i = √−1. The product of like signs is + if we count in units of 1 and − if we count in units of i.

Complex numbers count in both units, which we do not need for onions. Current nomenclature is to write complex numbers like this:

$$x + iy$$

where **i** is $+\sqrt{-1}$ and x counts in units of +1. The $\sqrt{+1}$ is omitted, being understood to be the conventional unit of counting and called *real*.

> Historical: It was shown by Argand in 1813 that complex numbers behave like two dimensional vectors that can be manipulated algebraically as ordinary numbers. Extensive use is made of this way of describing 2-vectors like alternating current in engineering. Hamilton extended the use of the roots of −1 to three dimensional geometry, called quaternions. Gibbs showed that 3x3 matrices were easier to use for this purpose, relegating quaternions to history. Notwithstanding the grade school simplicity of matrices (as long as you do not have to invert them), the resistance to their use in such fields as surveying and astronomy arose from the fact although geometric matrices have only 3 degrees of freedom, the computation of the of the 9 elements of the matrices and the scalar products using them, involve mixtures of addition and multiplication. This was very burdensome in the days of hand calculation when logarithm tables were used for high accuracy multiplication. This objection was greatly diminished by the introduction of calculating machines around 1850 and completely removed by electronic computers.

Functions of a complex variable

Any function we think of as being a function of a real variable, without even thinking that it is real, can also be a function of a complex variable. The function generally will also be complex, but it can be real: Take the simple case $y = x^2$ when $y = -1$. This can only be possible if $x = i$. Wavefunctions in quantum mechanics generally are complex.

Functions of a complex variable can be very different from their real counterpart. Of paramount importance in the quantum mechanics and the most famous example is this: What is e^{ix}? This is easily found by substituting ix for x everywhere in the above series. This is what you get:

$e^{ix} = 1 + i.x/(1) - x^2/(1.2) - i.x^3/(1.2.3) + x^4/(1.2.3.4) + i.x^5/(1.2.3.4.5) - x^6/(1.2.3.4.5.6)$ . . .

which if you know the series for sines and cosines is:

$$e^{ix} = \cos(x) + i.\sin(x)$$

The cosine series in in **black ink** and he sine series in gray ink.

which is why **i** makes waves and why wave mechanics needs **i**. It turns the least wavy function e^x, called exp(x) in spreadsheets and computer code, into the cosine(x) and sine(x), who share first prize for being the waviest functions, making as they do, waves forever.

Here is the family portrait:

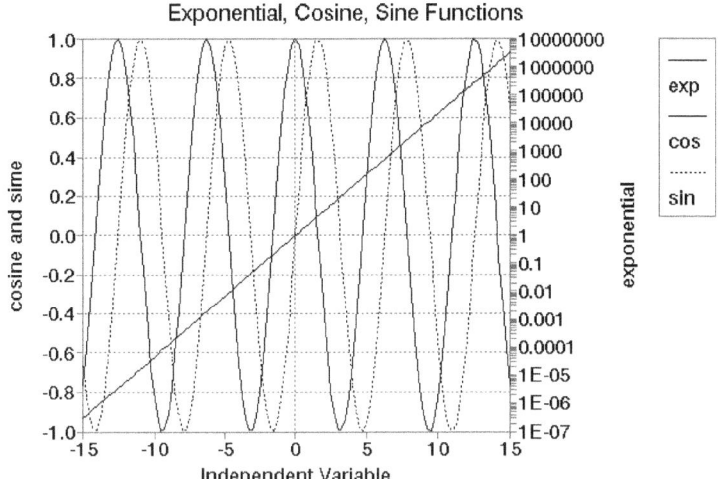

What a difference an i makes

In just a few waves of the sine and cosine function between -1 and 1 the exponential function rose from one ten millionth to ten million. I had to use a logarithmic scale or sine and cosine would have been squashed flat.

If you differentiate cos(x) twice you get -cox(x)

If you differentiate sin(x) twice you get -sin(x)

You can see this by differentiating the above black and gray series twice and getting the same series with the sign reversed.

Although they can get their fingernails dirty if not get their fingers burned, I hold mathematicians in great respect for their dedication to the study of such eternal truths. Napier didn't invent **e**, he discovered it.

Proof: There is many a planet in the universe where life is advanced beyond us. They all know:

e= 2.71828182845904523536028747135266249775724709369995...., but none of them, except on planet earth, ever heard of Napier.

271

Appendix E: Linear Algebra

Linear Algebra

The mathematical core of the quantum mechanics is linear algebra. When some list of numbers, (ξ, η, ζ) in the example below depend on some other list of numbers (x, y, z) in this example, mathematical life cannot get much simpler than this:

- each number in one list being simply proportional to each number in the other list
- the proportional terms simply add together, like this:

```
a.x + b.y + c.z =  ξ
d.x + e.y + f.z =  η
g.x + h.y + k.z =  ζ
```

where . is multiply and + is add, and where the constants of proportionality are a, b, c, d.... in this example. That's linear algebra. The use of Greek letters is just to get more wriggle room: It has no deep inner meaning except to high school kids who show superiority over their grade school siblings who cannot write them. After substituting relevant numbers for the symbols, the problem of calculating

(ξ, η, ζ), given (x, y, z)

reduces to elementary grade school arithmetic. The problem of calculating (x, y, z), given (ξ, η, ζ), reduces to ~~elementary~~ advanced grad$e$ school ~~arithmetic~~ mathematics, worse even than the Chinese ring puzzle.

Vectors

The arrays of numbers in the above equations:

(ξ, η, ζ),
(x, y, z),
(a, b, c), (d, e, f) and (g, h, k)

are called vectors, a term that comes from geometry where such a list is needed to specify the location of a point in three dimensional space.

When only a single number is needed, such as the temperature at that point, it is called a scalar. However, in the quantum mechanics vectors may have 2, or 3, or more than 3 or even an infinite number of elements. If infinite, most of the numbers are so small they can be ignored in practice.

Scalar products

Each equation in the previous section above defines a *scalar product* of two vectors. For example, the first equation is the scalar product of the vector **(a,b,c)** and the vector **(x,y,z)**. It is called scalar because the result is not a vector but a simple number, in this case ξ. The next two equations give the scalars η and ζ. However, taken together as:

$$(ξ, η, ζ)$$

is a vector.

In the geometric example, **(a,b,c)**, **(d,e,f)** and **(g,h,k)** are called unit vectors because $a^2+b^2+c^2=1$, etc. **a,b,c,d,...** really have no dimensions like microns, millimeters, meters or light years, but are dimensionless ratios: Geometrically they are the cosines of the angles between the vector and each of the orthogonal reference axes because they are the projections of the vector onto each axis. In geometry they are called direction cosines.

Going with the above example, **x, y, z** are the coordinates of some vector relative to a set of three orthogonal axes reference frame, call it the "old X,Y,Z" system, and **(ξ, η, ζ)** the coordinates of the same vector relative to a "new" system with 3 orthogonal axes. Then **(a,b,c), (d,e,f)** and **(g,h,k)** are the three unit vectors orthogonal to one another that define the "new" reference frame in terms of the "old": The numbers **a, b, c, d, ...** refer to the old system. The three linear equations above are called the *inner, scalar or dot products* of the vectors

(a,b,c) and (x,y,z) written (a,b,c).(x,y,z) = ξ, the projection of x,y,z vector onto the a,b,c axis.

(d,e,f) and (x,y,z) written (d,e,f).(x,y,z) = η the projection of x,y,z vector onto the d,e,f axis.

(g,h,k) and (x,y,z) written (g,h,k).(x,y,z) = ζ the projection of x,y,z vector onto the g,h,k axis.

Why does

$$a.x + b.y + c.z = ξ \ ?$$

The reason for this is that the (x ,y, z) vectors projection onto the X axis is x, and a vector of 1 projects onto the new X axis vector (a, b, c) as a, so the x component projects the fraction a.x onto the new X axis = the Ξ. axis But the y component of the (x ,y, z) projects b.y also onto the new Ξ axis and the z component projects c.z onto the new Ξ axis, so we have to add all these component to get the total new Ξ component. ξ could be zero if all components canceled because they can can be positive or negative. This whole mishigas has to be repeated for η and ζ, which is why matrices were hated at first. But it is grade school arithmetic and now your computer does it for you.

If you can point at something in the sky, representing a vector, namely your pointing arm and finger, pointing by three orthogonal components may seem artificially abstract. Here is a counter example. An airplane pilot knows somehow that to execute a certain maneuver, a torque must be applied to the airplane about an axis the pilot designates by pointing a finger at some point in the sky and prays to the gods for help. It just so happens that there is a giant hex head bolt welded to the airframe that points in the same direction. The gods reach out of the sky with a torque wrench and apply the required torque to the bolt until the pilot signals them to stop. This torque is a vector along the axis of rotation with some understood convention like clockwise is positive. There is a better way to do this and it is precisely equivalent. Suppose the maneuver is to turn right. The pilot applies right aileron which applies a torque to the airplane about the longitudinal X axis. This lifts the left wing and lowers the right wing (roll). With the airplane tilted, there is a centripetal force pushing the airplane to the right. But the pilot now must apply right rudder to keep the X axis tangent to the desired circle by rotating the airplane about a vertical Z axis (yaw). Now the aircraft is rotating in a tilted plane which causes the nose to fall below the horizon, so the pilot applies up elevator to rotate the airframe about a horizontal Y axis from wing tip to wing tip (pitch) to hold the nose on the horizon. The sum of these three torques is identical physically to one applied about some single axis. It is in fact easier for the pilot to visualize what is needed by thinking of the orthogonal components of torque than thinking in terms of just one vector that is their sum.

Matrices

The above equations are written in this shorthand notation, where the arrays in ‖ brackets are called matrices:

$$M * X = \Xi$$

or

$$\begin{Vmatrix} a & b & c \\ d & e & f \\ g & h & k \end{Vmatrix} * \begin{Vmatrix} x \\ y \\ z \end{Vmatrix} = \begin{Vmatrix} \xi \\ \eta \\ \zeta \end{Vmatrix}$$

This is shorthand for the equations of linear algebra:

$$a.x + b.y + c.z = \xi$$
$$d.x + e.y + f.z = \eta$$
$$g.x + h.y + k.z = \zeta$$

The properties of the linear equations depend only on the array **M** as an *operator* and not on the vector **X** on which it operates. **M** is called a matrix. Don't feel bad if you didn't know this. Heisenberg didn't either. Born did.

Suppose:

matrix **G** operates on vector **x** to give vector **y**, then

matrix **H** operates on vector **y** to give vector **z**

We can write this as **G.H.x = z**, leapfrogging **y**.

Equivalent to this is:

J.x = z where **J** = **G.H** each element of C in row i and column j is the scalar product of row n of **G** with column m of **H**.

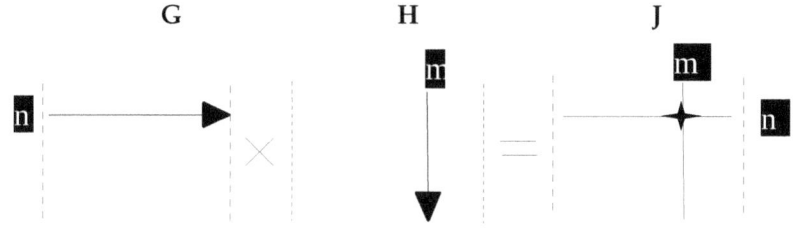

Written **G.H=J** and called *matrix multiplication* because it obeys all of the algebraic laws of ordinary multiplication with this exception: **G.H = H.G** is not generally true. It is only true if **G** and **H** have the same eigenvectors, discussed later. Matrix multiplication is what Born recognized Heisenberg was doing.

Given a matrix equation like the one above:

$$M * X = \Xi$$

if you are given M and X you can calculate Ξ using the rule given above in which the index n = 1 for the matrices H and J, given X it is easy to calculate Ξ, but a job best left to a computer.

But suppose you are given M as before but you have to find X given. One may guess that it is also a linear algebraic problem like this:

$$W * \Xi = X$$

W, called the inverse matrix of **M**, can be calculated from M, but is not only a job for a computer because of the mountain of tedious high school algebra involved, but also a job for a mathematician because of the pitfalls that await the fool who rushes in.

Matrix algebra shows us simple and elegant ways of understanding this without getting our fingernails dirty doing the arithmetic. Multiply the first equation above by W:

$$W * M * X = W * \Xi, \text{ which is just } = X$$

This can only be true if **W*M = 1.**

But **W**, **M** and **1** must be matrices, so what is **1**? It is called the unit matrix, because it does nothing to any vector it multiplies, just like multiplying by 1 in arithmetic:

$$\begin{Vmatrix} 1 & 0 & 0 \\ 0 & 1 & 0 \\ 0 & 0 & 1 \end{Vmatrix}$$

is a 3x3 example. If your spreadsheet program inverts **M** to get **W** and doesn't get this unit matrix when you multiple **W*M**, get a better spreadsheet.

Appendix F: Differential Equations

Most of the laws of physics are simple when they appear as *differential equations* which are relationships between functions and their derivatives. They show how the function and its rates of change are related. They do not tell us what we want to know, because the function, call it f, in the differential equation is unknown – the challenge is to find out what it is. Finding out say how f changes with time is called solving the differential equation and doing this, especially with partial differential equations, is often very difficult. Because the moon can be seen from anywhere most of the time, it can be used as a clock to show standard time, allowing the determination of longitude of any point on earth and of ships at sea. But you had to know the moon's position as a function of standard time. However, it took a major well funded international effort 300 years to solve the differential equations of the earth, moon, sun system. The differential equations were simple: Newton's second law force = mass . acceleration. Let's write this in the above notation: Let the position of some mass point be r and its velocity v. Its acceleration is dv/dt. Calling force f and mass m, then Newton II is

$$f = m.dv/dt$$

But v is the rate of change of position r or dr/dt. So f = m.d/dt(dr/dt). Here d/dt is the differentiation operator operating on itself operating on r. Clumsy. Leibniz agreed and said let's write d/dt(d/dt) as: d^2/dt^2. So this operator operating on r is written d^2r/dt^2 which is the acceleration in r. So instead of clumsy f = m.d/dt(dr/dt), we can write:

$$f = m.d^2r/dt^2$$

This is Newton's second law. If you have a mathematical expression for f, you've got yourself a differential equation for **r** and can find it as a function of the time. You have to know **r** and the **r** momentum at some instant of time, which Heisenberg says is impossible. Good luck. It may not be easy, even if you don't rum into Werner. To show that it is possible we will set up and solve a differential equation after the next subchapter.

Linear operators

What linear means is that it some operation say **A,** like the differentiation operators above, acts on the sum of functions say f, g, h.. the result of **A** operating on f+g+h+.., written as **A**(f+g+h+..,) is also **A**f+**A**g+**A**h+.., It does not matter if you add the functions together first then operate on the sum, or operate on each one separately then add the sum of the results. An example of a linear operator is multiplication. If you have n teams each with b boys and g girls, the total number of players is n(b+g) = nb + ng

An example of an operation that is <u>not</u> linear is squaring a sum of numbers:

$(f+g+h)^2$ is <u>not</u> $(f^2+g^2+h^2)$. Most operations are not linear.

The differentiation and integration operators:

d/dx, d^2/dx^2, $\partial/\partial x$, $\partial^2/\partial x^2$ etc and $\int dx$

are all linear operators. It does not matter it you differentiate a sum of a bunch of functions. You get the same result if you differentiate each separately and add the results. Same with integration. It follows that a linear operation performed on a constant times a function is the constant times the operation performed on the function.

Demystifying differential equations

We will derive a differential wave equation wave from scratch and solve it. We are using Newtonian mechanics to solve the wave equation of a stringed instrument because of the strong parallels with quantum mechanics wave functions. The math jargon arose in connection with these equations of classical physics. The solution will pop out as eigenfunctions, and its them we are after.

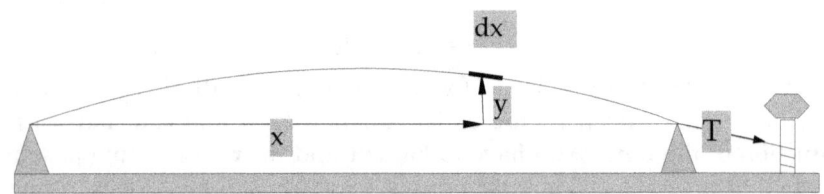

The string stretches from the bridge to the tuning pin fret and is under tension T. With the string vibrating we take a snapshot showing at distance x along the string it is displaced from its silent position by y. Consider a short length dx of the string at that point, having a finite length but it is really the infinitesimal dx.

If the string is vibrating and we look at a snapshot near **dx** shown as a **bold** line, we may see something like this, where the string is straight at dx:

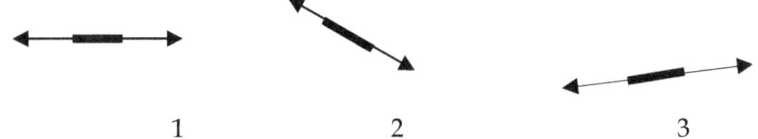

1　　　　　　　　2　　　　　　　　3

The arrows represent the force of tension in the string. Call it T. Newton III tells us the force on each end of dx will be T: equal and opposing forces. There is no net force on dx because it is being pulled equally in opposite directions. So there is no upward or downward force either as long as the string is straight at dx.

But this is not true if the element dx **is** bent into a curve like this:

It does not matter which case 1, 2 or 3 above that we are looking at, so we look at case 1 which makes the math easier. What's pulling dx up is the amount the slope of the string changed over the distance dx multiplied by T. We are assuming the string is very flexible. Using Leibnitz' notation, and calling the vertical displacement of the string y at distance x along the string where dx is located, the rate of change of the slope is:

$$\partial^2 y / \partial x^2$$

And the actual change of slope is;

$$(\partial^2 y / \partial x^2) dx$$

So the upward force on **dx** is:

$$T(\partial^2 y / \partial x^2) dx$$

It looks like Newton III is being violated because there is no equal and opposing upward force. There <u>is</u> an equal and opposing force because of Newton II. It is the inertial force of the string's mass that resists being accelerated. Suppose the string has a mass s per unit length. So the mass of dx is just s.dx. This mass is resisting being accelerated which supplies the opposing force. In piano low registers the steel strings are weighted by winding brass wire around them to increase s without making the string so stiff as to make for a twangy sound due to the stiffness of the wire which would add new forces inside the wire making it vibrate more like a gong. We can now use Newton II, Force = Mass x Acceleration, to write the equation of motion of the string:

$$T(\partial^2 y/\partial x^2)dx = s.dx.(\partial^2 y/\partial t^2)$$

Because, using Leibnitz notation

$$\partial^2 y/\partial t^2$$

is the acceleration of in the y direction being as it is the time rate of change of the y velocity with time t: Velocity in y is:

$$\partial y/\partial t$$

When dx is infinitesimal we are talking about what happens at any point x along the string measured from the fret. dx is infinitesimal but it is not zero, so we can cancel dx giving the differential equation of motion of the string:

$$T(\partial^2 y/\partial x^2) = s.(\partial^2 y/\partial t^2)$$

or grouping the constants s and T:

$$\partial^2 y/\partial x^2 = (s/T).\partial^2 y/\partial t^2$$

This is the differential equation of motion of the string. It only tells you how at some point at x along the string where it is displaced y how that point is accelerating due to the curvature of the string at that point. Otherwise we have no idea what y is, or even what the velocity $\partial y/\partial t$ is. To find out how y changes with x and with time t, have to solve the equation. A tried and true way to do this is to guess. We are really doing the integration operation which should be left to mathematicians, since we are going from derivatives to functions. But our guess will be right! (Because I already know the answer).

The string is anchored at each end where y = 0, always. If the length of the string is π = 3.1415626536...., we can try y = Sin(x) which is always 0 when x=0 and x=π. If the string is not π long we can still use the same equation but have to introduce another constant that makes the algebra messier and no more instructive. (You cannot be this sloppy if you have to do this for a living). But we assume the length of the string is π. (I am retired). If you differentiate Sin(x) twice you get −Sin(x). To see why read appendix D. But the string is not going to remain frozen in space. It is going to vibrate up and down starting say at maximum positive, down to 0 to maximum negative then back again. Looks a lot like Cosine. But we don't know how fast so we will guess Cos(kt). We don't know what k is but hope to find out. So we guess the solution is:

$$y = Sin(x).Cos(kt)$$

If this is not a solution to the above differential equation, we will get into trouble.

Substituting it into the differential equation gives:

$$(\partial^2 \text{Sin}(x)/\partial x^2).\text{Cos}(t) = -\text{Sin}(x).\text{Cos}(kt) = (s/T).\partial^2 y/\partial t^2$$

remembering that partial differentiation with respect to x on the left treats any function of the other independent variable t as constant. And differentiating Sin(x) twice with respect gives −Sin(x).

Now we do the right hand side:

$$-\text{Sin}(x).\text{Cos}(kt) = (s/T).\partial^2 y/\partial t^2 = (s/T).\partial^2[\text{Sin}(x).\text{Cos}(kt)]/\partial t^2$$

Treating Sin(x) as a constant with respect to t, when we differentiate Cos(kt) with respect to t.

$$-\text{Sin}(x).\text{Cos}(kt) = (s/T).k^2[-\text{Sin}(x).\text{Cos}(kt)]$$

Differentiating Cos(kt) twice with respect the variable kt gives −Cos(kt), but we are supposed to differentiate with respect to t, not kt. We correct this by multiplying by $d(kt)/dt = k$ each time, which is where k^2 came from. (Do not ask why, or I will suggest you take a calculus course). The Sin and Cos functions generally are not 0 so we can cancel them getting:

$$k^2 = (T/s)$$

or

$$k = \sqrt{(T/s)}$$

k is the frequency in radians per second. We have to divide it by 2π to get cycles per second or hertz. The pitch of the note is the nominal pitch, say middle C and is called the fundamental.

So the guess

$$y = \text{Sin}(x).\text{Cos}(kt)$$

is a solution to the equation only if $k = \sqrt{(T/s)}$ which is the pitch of the note the string will resonate.

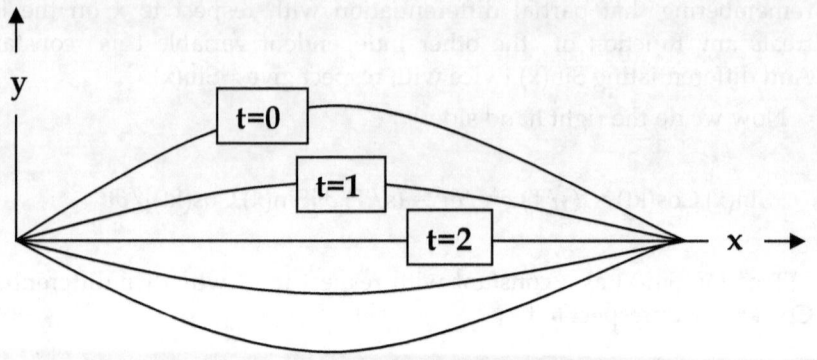

Suppose instead you had guessed:
$$y = \operatorname{Sin}(2x) \cdot \operatorname{Cos}(jt)$$
because $\operatorname{Sin}(2x)$ is also 0 at the bridge and tuning pin fret.

It is a solution to the equation only if $j = 2\sqrt{(T/s)} = 2k$ which is the pitch of the note the string will resonate at an octave higher than the above resonance. Is this really possible? Indeed if you damp a guitar string at its exact center with a finger tip and pluck it somewhere else, it will resonate like this:

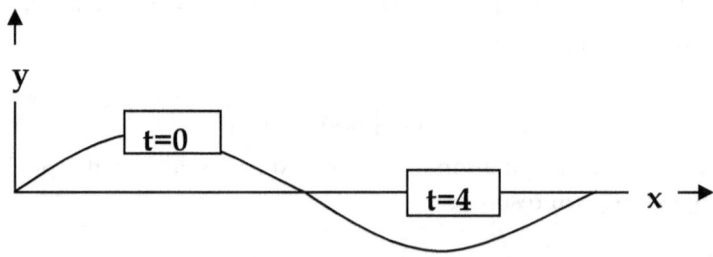

And the pitch will be an octave higher than the fundamental.

And here is another: **Sin(3x).Sin(3kt)**

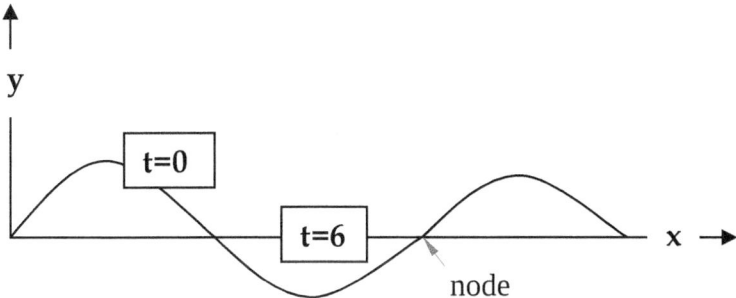

If you damp a guitar string exactly 1/3 the way from the fret, a node, and pluck it, it will resonate like this. This corresponds to the 7th key above the next octave.

Musicians call these the harmonics of the pitch note which is called the fundamental in physics.

You guessed it. This can go on forever. Mathematically there are infinitely many harmonics all of which are solutions to the wave equation.

$y = Sin(1x).Cos(1kt) = E_1$
$y = Sin(2x).Cos(2kt) = E_2$
$y = Sin(3x).Cos(3kt) = E_3$
$y = Sin(4x).Cos(4kt) = E_4$
$y = Sin(5x).Cos(5kt) = E_5$
$y = Sin(6x).Cos(6kt) = E_6$

And so on forever are all solutions to the differential equation:

$$\partial^2 y/\partial x^2 = (s/T).\partial^2 y/\partial t^2$$

where k = $\sqrt{(T/s)}$ and k/(2p) is the fundamental pitch in hertz (cycles/second), T is the tension in the string and s the mass/unit length of the string.

With the exception of the 7th and 11th harmonics, all harmonics are fairly accurately represented on the equal tempered scale so sound like notes you can play on the keyboard. In pianos, the hammer strikes the string is 1/7 of the way from the tuning pin fret which is a node, so killing the nasty 7th harmonic, makes the sound more mellow. The 7 th harmonic on a violin is not discordant, on a piano it would be because of the equal tempered scale. A Humpback whale may find the sound of a piano as unendurable as bagpipes are to those without Celtic blood.

Eigenfunctions

Mathematicians call these the harmonics *eigenfunctions* of the differential equation. They are the *characteristic functions* of differential equations that are linear, but even in English the name eigen.. stuck. The pitches of each harmonic is called its *eigenvalue*.

Any linear combination of eigenfunctions is a solution to a differential equation that has the property of being **linear**, which our equation is. For example:

$y = 0.4E_1 - 0.2E_3 + 0.3E_5$

is a also a solution because the eigenfunctions $y = E_1, E_3$ and E_5 are solutions.

The negative coefficient simply reverses the polarity of negative and positive y.

We are getting close to explaining the math of the quantum mechanics. We have to look at properties of eigenfunctions or eigenstates as are they are also called when we are looking at quantum phenomena like polarization and spin of electrons that are not described using wave functions. We can then look at what Heisenberg did with his linear algebra. The equivalence of these two approaches was known to the mathematician Hilbert from his study of differential equations, like the above piano musical instrument string problem, before the days of the quantum mechanics. The eigenfunctions have a property that is fundamental to the interpretation of the quantum mechanics. Eigenfunctions are orthogonal to one another.

Eigenfunctions are orthonormal

In Appendix E we described a fundamental operation of linear algebra, the *scalar product* or *inner product* of two vectors. If A and X are vectors, it is sometimes written A.X and called the *dot product* in geometry. If A has components a,b,x and X has components x,y,z, their scalar product is a.x + b.y + c.z where the "." means multiplication. In this example, the vectors have only 3 components but they may have any number including infinitely many. If the vectors are also unit vectors they have the property that their scalar product is 1 if they point in the same direction (think geometry), 0 if they are orthogonal and -1 if they point in opposite directions. We will describe as a *basis* a set of orthonormal unit vectors that are all are orthogonal to one another. In other words the scalar product of any two is zero, but the scalar product of any one with itself is 1. An example may be a unit vector pointing up, one pointing east and one pointing north.

What does this have to do with eigenfunctions? Take the above eigenfunctions for a piano and tabulate them in two columns for say 16 equally spaced points labeled 0 to 16 and for angles of p.(point number)/16. At the head of each column put in a harmonic number. You can include the time term but, I will assume below that we hold the time at zero so all these cosine terms are 1. In a third column put in the product of each of the harmonic functions and at the top of that column the sum of the column. This is just the scalar product of the two eigenfunctions, which you can also do with the scalar product function of a spreadsheet called the sumproduct(A,B). If you enter the same harmonic number in each column the scalar product will be 8 in this example but you can make it come out as 1 by using a normalizing constant of $\sqrt{8}$ in each harmonic. If you key in different harmonic numbers for each harmonic, the scalar product will be zero. This is called the delta function. If the column index is i and the row index is j. The function is written:

$$\delta(i,j) = 1 \text{ if } i=j \text{ and } \delta(i,j) \neq 0 \text{ if } i = j$$

Technical note: Since the functions are continuous, the strict definition of the scalar product is $\int \sin(i.x).\sin(j.x).dx$. Because the functions being integrated are periodic, the integral is given exactly by the discrete scalar product provided one keeps the harmonics numbers under the number of tabular values that represent the range 0 to π which otherwise causes aliasing.

Linear superposition of eigenfunctions

Suppose the piano string is deformed at time 0 into any waveform y(x) that is 0 at the bridge and tuning pin fret, and we want to describe this as a linear superposition of eigenfunctions $E_1, E_2, E_3, E_4, E_5, E_6, ...$
Write
$$y(x) = a.E_1 + b.E_2 + c.E_3 + d.E_4 + e.E_5 + f.E_6 \ldots .$$
Suppose we want to find the coefficient d that multiplies E_4
Take the scalar product of both sides of the equation with E_4. Every

term on the right side is zero except the scalar product of E_4 with E_4 which is 1, leaving only d on the right. On the left is $<y(x)|E_4>$ using Dirac's notation. We can evaluate it and get d. Obviously we can do the same thing for every eigenfunction and get a,b,c,e,f,....as well.

Hilbert Space and the quantum mechanics

It may seem that vectors in Hilbert Space of infinitely many dimensions is a heckuva roundabout way of describing a piano string. If you take a high speed camera or strobe lamp and photograph the string in resonance, you see $y(x)$. Why the eigenvector mishigas?

Here is the reason for the quantum eigenvector story: If you observe a system quantum mechanically that is in predicted state $y(x)$, you do not see $y(x)$. You see one of 'its' eigenstates E. Specifically, if you observe it with an operator **A**, you will find the system in one of the eigenstates of **A**. If you repeat the experiment, like as not you will see it in a different eigenstate.

Everett showed us a simpler explanation: All the states of awareness are equally real but are exclusive. He did not use the word real, but his analysis of the quantum math is made clearer if one sees every state as experienced in consciousness. The exclusivity is that conscious experience of each eigenfunction does not bind in awareness. This would not be possible if consciousness were in the physical world as von Neumann proclaimed. If it were, inconsistent states would be caused by co-existing inconsistent physical world elements. Everett did use the word *aware*. If you repeat the same experiment after very little elapsed time – too short for change to be significant in the equation:

$$\partial\psi/\partial t = \mathbf{A}\psi$$

your experience of the same eigenstate will bind with the one you remember the last time. All of this follows from the math of the above equation.

The uncertainty of Heisenberg and Dirac formalized by von Neumann is unnecessary metaphysics.

<u>The reason for this exercise is to show that the eigenfunctions appear as natural solutions to the differential equation of motion. This is how Schrödinger got quantization into the quantum mechanics of hydrogen. I did not use his equation in this exercise because the math is too complicated.</u>

Appendix G: The Twin Paradox of the Special Theory

This paradox is usually presented as one of pair of twins goes on a fast journey leaving her other twin at home, which is a vast system of relatively stationary instruments of metrology, mainly clocks . When the wayward twin, who only carries a wrist watch, returns to rejoin her twin sister, the difference in age between them is shocking. Without getting into the quantum mechanics, this property of reality, which is seen in elementary particle physics where the lifetimes are of the order of a microsecond, is the most disruptive, easily understood notion of physical reality.

To my high-school cohorts I am indebted for the illustration of this paradox given below. It has the merit that it describes entire systems of metrology in relative motion, but the demerit of requiring sextuplets, instead of twins. Commuter trains from different directions converged onto parallel adjacent tracks in Braamfontein for the final run to Park station in Johannesburg, their common destination. Sometimes, when the train speeds matched, schoolboys would fling open passenger doors and leap from one train to the other. They did this for the same compelling reason people climb Mt. Everest: Because it is possible.

I never saw a schoolgirl stupid enough to do this, so I will invent her: She's Ann, on an intergalactic train going North, or South – does not matter. The cars have no wheels. There are no tracks – they are space cars not rail cars: Each car is a light-year long. Ann is traveling with Beth, one of her identical sextuplet sisters. They are both 17 and are talking about their identical sister Claire riding on the same train twenty cars North. Ann notices the train to the West, an easy leap away, sometimes runs at the same speed then speeds up to nearly the speed of light, headed in Claire's direction, until 20 cars have passed, one per minute, then slows down to the same speed as her train. The Principle of Relativity holds that what is observable only depends on the relative velocities of the trains, so I left out the train tracks to keep ones mind from latching on to them; we are not interested in "how fast" Ann and Beth's train is going "through space" or even "which direction".

Ann decides to visit Claire and leaps onto the West train when the speeds match. When it speeds up by nearly the speed of light relative to the train she leapt off, people on that train see her as frozen in time: her heart is beating only 60 beats a year. But she sees them the same way. Then she notices as she as she looks through the window in each car of the train she left, the dates on their I-pads are a year later from one car to the next, minute by minute. She is traveling down their time dimension. Such travel is always into their future for quantum mechanical reasons described in the chapter *The Arrow of Time*. These reasons are obscured by conventional quantum mechanics metaphysics. When the trains match speed again Ann leaps into Claire's car and visits with her.

After a cup of tea, Ann says "Bye Claire. Gotta go" and leaps onto the train on the East side because she knows it goes South and will take her back to Beth's car. When it speeds up to nearly the speed of light, the same thing happens again: People on the train she was on see her as frozen in time: her heart is beating only 60 beats a year. But she sees them the same way. Then she notices as she looks through the window of each car of the train she left, the dates on their I-pads are a year later in each car she passes, minute by minute. Such travel is always into their future. When Ann's car pulls alongside Beth's, Ann leaps back and rejoins Beth. Ann has been gone about an hour. Beth hasn't seen Ann in 40 years. Beth is now 57. So is Claire, twenty light years away. Ann is still 17. In the hour Ann was away, she traveled 40 years into Beth and Claire's future.

Beth is furious. She wants to be 17 again, like Ann. To undo the mess Ann has made, Beth leaps onto a South bound train for twenty minutes, leaps back, then leaps onto a North bound train, returns to Ann and leaps back. Now she is the same age as Ann again, but she does not get what she bargained for: Both Ann and Beth are now 57.

Suppose before Ann makes her first leap, she and Beth see their identical sisters Dana and Elsa on the West train. Ann and Dana both jump trains, trading seats. Dana wants to go back to visit sister Fay 20 cars South on the train she was on. But the train Dana is on, is not going to accelerate. Fay's train, the one Dana was on and now Ann is on, will accelerate.

It is commonly claimed that the Special Theory makes travel faster than light impossible. This claim is false. Velocity is distance divided by time. When Ann goes to visit Claire, she is interested in going 20 light years in Claire's space divided by the time elapsed on Ann's own watch, that is 20 minutes. She is traveling 500,000 times faster that the speed of light in Claire's space, the only distance that matters to her: the distance of 20 light years to Clair. The only time that matters to her: the travel time of 20 minutes on her watch it takes her to get there. This is exactly what you do when you calculate the speed you drove to Los Angeles. Both Claire and Ann agree that the relative speed of the trains is a bit less than the speed of light but each sees the other train cars as only one light minute long. Claire is 37 when Ann arrives. The cause of this paradox is belief in the ultimate reality of the Physical World in which such inconsistency is absurd. The journeys Ann and Beth take are journeys in consciousness. The speed of light limitation is only valid if the measuring rod and clock belong to the same system, i.e. they are relatively motionless.

Accelerating people to a relative speed close to that of the speed of light is practically impossible, but these time dilation effects are commonly observed with elementary particles which can so be accelerated. A fast meson can travel much further into our future than its natural lifetime of about 2 microseconds. Bulk matter can be accelerated to speeds comparable to the speed of light – it happens in supernova jets.

It may seem that such time travel is only speculation since no one will ever be accelerated to such speed. It has in fact been demonstrated by human beings riding in an airplane: Not by counting heart beats or looking at age wrinkles, but by looking at atomic clocks they took with them which are precise to a nanosecond. This experiment is done to greater accuracy with atomic clocks on artificial satellites, which move at 5 miles a second or 8 km/s. The GPS satellites carry atomic clocks that emit pulses every second at the same time in the reference frame of the surface of the earth. The arrival times at a GPS device can tell its position to 1 meter which requires timing accuracy of 1/300 microsecond or 3 nanoseconds. The GPS system would not work if the Special Theory and General Theory corrections were not applied. It would be grossly incorrect.

The twin paradox as it is called is usually described as one twin making a journey relatively to the other who remains in a fixed in a space through which the other twin travels. The advantage of the parallel train model it that one can see how it works when two identical spatially extensive systems are in relative motion. We have to ask how to accelerate an entire train to near the speed of light. The obvious answer is that all the cars must start accelerating at "the same time". But these words have no absolute meaning: The clocks of any system can be synchronized, but unless you are moving relative to that system at any velocity v for which the approximation v^2 = zero is accurate, where v is the velocity relative to that of light, their clocks will not appear to be synchronous – you will not only be traveling through their space but also through their time.

We look at Dana's journey to visit Fay. We suppose all the cars of the train Ann, Elsa and Fay are on, call it train W, start moving at the same time in the system of the train Beth, Dana and Claire are on: Call theirs train X, and time $t_X = 0$ in clocks synchronous in X. Fay, at the back of her train, and Ann and Elsa at the front, will accelerate at the same time Beth and Dana see the front of the W train accelerate. Fay will pass the full length of train X in 20 of her train W minutes, joining Beth and Dana when Fay travels the 20 light years of train X at near the speed of light taking 20 years. Beth and Dana will be 37.

But disaster has struck train W. You can see why by looking in the X frame at it the instant it reaches the speed of light. All cars of train W will be in one to one correspondence with the cars of train X, but train X will see a 20 light year measuring rod on train W to be only 20 light minutes long. Train W will be torn asunder.

We try something else. The locomotive in front of train W stars pulling it at $t_X = 0$. The train is infinitely rigid but causality cannot travel faster than the speed of light, so it will be 20 years in X time before Fay starts moving and she, Dana and Beth will be 37. Fay gets to them 20 minutes later, but Dana and Beth are now 57. We can avoid this premature aging problem by having the acceleration of train W start 20 years earlier, before the sextuplet sisters are born. Three years later they will be born and grow up going to school together. When they are 17 and graduate from high school, for a graduation present they are given an intergalactic train trip and are transported to where we want them at the speed of light where they arrive at $t_X = 0$ when they are all still 17. Now Fay starts accelerating on train W at that time and joins Dana and Beth when they are 37. Fay is 17. But doing this is an even worse disaster to train W. You can see why by looking at what happens in just one year of X time. The locomotive will be 1 light year North of the front of train X and acceleration will be beginning 1 light year south. The front car will be 2 light years long in the reference frame of train X and it should be 1 light minute. Train W will be torn double asunder.

So we try something else. We switch the locomotive and caboose of train W and have it start pushing the train at $t_X = 0$. The train is infinitely rigid but causality cannot travel faster than the speed of light, so it will be 20 years in X time before Beth and Dana see the caboose, now on the front of train W, accelerate. They will be 37 and 20 minutes later Fay will arrive, still 17. It may seem that this will also wreck train W because the back of the train will be moving forward at nearly the speed of light into stationary cars. But the point of acceleration will be moving faster so a crunch will not happen. After 1 year in X time, the moving part of train W will be 1 light minute long appended to 19 light years of stationary cars. After n years in X time, the moving part of train W will be n light minutes long appended to (20−n) light years of stationary cars. The coupling point between moving and stationary in frame X will be moving North at the speed of light. When Beth and Dana see the caboose accelerate, the whole W train will be in relative motion and it will be 20 X light minutes long in the X train reference frame. If you are moving at the speed of light with the point of acceleration it will travel the length of train W instantaneously: The whole train will accelerate at the same time in a reference frame moving at he speed of light. Classical physics has causality traveling at infinite velocity so the whole W train accelerates at the same time in any reference frame since all are equivalent, but classical physics is wrong.

In all cases discussed above, the person who was on the train that accelerated traveled forward in time in the frame of those on a train that did not accelerate. It may seem that time dilation was caused by acceleration. This is not true: Experimentally acceleration has no such effect. Time dilation is caused by changing reference frames and staying in the new frame for some time during which one travels through both the space and time of the other frame. You cannot drive from San Francisco to Los Angeles without accelerating, but doing so only gets you to the end of the I5 on-ramp. You have to stay on I5 for 5 hours to get to LA. At 60 miles/hour your travel into the LA future would not be measurable even with a nanosecond accuracy atomic clock.

The penny pulp drama of the above scenario should not obscure the correct kinematics of reality. There is no place in the physical world of common sense for these kinematics. We have only two choices:

1. delusion or
2. 2. dump the "physical world" as we know it.

Appendix H: The Cast

1927 Solvay Conference on Quantum Mechanics

Photograph by Benjamin Couprie, Institut International de Physique Solvay, Brussels, Belgium.

Names appearing in this book are shown in **bold** font.

Auguste Piccard, Émile Henriot, Paul Ehrenfest, Édouard Herzen, Théophile de Donder, **Erwin Schrödinger**, Jules-Émile Verschaffelt, **Wolfgang Pauli**, **Werner Heisenberg**, Ralph Howard Fowler, Léon Brillouin,

Peter Debye, Martin Knudsen, William Lawrence Bragg, Hendrik Anthony Kramers, **Paul Dirac**, **Arthur Compton**, **Louis de Broglie**, **Max Born**, **Niels Bohr**

Irving Langmuir, **Max Planck**, Marie Skłodowska Curie, **Hendrik Lorentz**, **Albert Einstein**, Paul Langevin, Charles-Eugène Guye, Charles Thomson Rees Wilson, Owen Willans Richardson

John von Neumann

Richard Feynman

Hugh Everett III

Leonard Mandel

David Hilbert

Kurt Gödel

Pascual Jordan

David Bohm

John Bell   Alain Aspect

Ernest Rutherford   Micheal Gazzaniga

www.ingramcontent.com/pod-product-compliance
Lightning Source LLC
Chambersburg PA
CBHW071411180526
45170CB00001B/72